Emerging Technologies in Vessels and Automated Drones for Maritime Innovation

Froilan D. Mobo
https://orcid.org/0000-0002-4531-8106
Philippine Merchant Marine Academy, Philippines

IGI Global
Scientific Publishing
Publishing Tomorrow's Research Today

Vice President of Editorial Melissa Wagner
Director of Acquisitions Mikaela Felty
Director of Book Development Jocelynn Hessler
Production Manager Mike Brehm
Cover Design Jose Rosado

Published in the United States of America by
IGI Global Scientific Publishing
701 East Chocolate Avenue
Hershey, PA, 17033, USA
Tel: 717-533-8845 | Fax: 717-533-7115
Website: https://www.igi-global.com E-mail: cust@igi-global.com

Library of Congress Cataloging-in-Publication Data

Names: Mobo, Froilan Delute, 1979- editor
Title: Emerging technologies in vessels and automated drones for maritime innovation / edited by, Froilan Mobo.
Description: Hershey, PA : IGI Global Scientific Publishing, [2026] | Includes bibliographical references and index. | Summary: "This edited book explores the transformative role of emerging technologies in the maritime sector, with a particular focus on automated drones, artificial intelligence, autonomous vessels, and smart navigation systems"-- Provided by publisher.
Identifiers: LCCN 2025021852 (print) | LCCN 2025021853 (ebook) | ISBN 9798337335469 hardcover | ISBN 9798337335476 paperback | ISBN 9798337335483 ebook
Subjects: LCSH: Shipping--Technological innovations | Ships--Technological innovations | Drone aircraft | Artificial intelligence--Industrial applications
Classification: LCC HE571 .E44 2026 (print) | LCC HE571 (ebook) | DDC 387.5--dc23/eng/20250730
LC record available at https://lccn.loc.gov/2025021852
LC ebook record available at https://lccn.loc.gov/2025021853

British Cataloguing in Publication Data
A Cataloguing in Publication record for this book is available from the British Library.

Table of Contents

Detailed Table of Contents

Maritime Search and rescue (MSAR) solutions must be intelligent, nimble, and autonomous due to the growing frequency and complexity of maritime emergencies, which are caused by climate change, growing coastal populations, and increased marine traffic. This study introduces a multi-sensor, AI-enabled unmanned aerial vehicle (UAV) system intended to improve maritime SAR operations in real time. To provide synchronized data collection, in-situ processing, and quick decision-making, the suggested framework combines thermal, RGB, and LiDAR sensors with a Pixhawk 4 flight controller and a Jetson Xavier NX edge computing module. The domain-specific AI4SAR dataset is used to train YOLOv5s, a compact deep learning architecture that is enhanced with both synthetic and real marine scenes to guarantee resilience in a range of environmental circumstances. Transfer learning, ONNX export, and TensorRT acceleration are used to optimize the model, allowing for inference speeds exceeding 30 frames per second with 92.1% precision, 88.7% recall, and 90.3% mAP@0.5. The outcomes highlight the system's applicability for use in actual SAR situations, where operational resilience and real-time item detection are essential for saving lives in hazardous maritime environments.

In this chapter, wc dcscribc an anomaly detection system for UAS telemetry data based on unsupervised Autoencoders. We created telemetry transmissions with encrypted communications using AES and ChaCha20 in Python, following the MAVLink protocol, to maintain a realistic communications environment. The encrypted logs were transformed to CSV files and pre-processed with normalization and feature selection to prepare the data for model training. An Autoencoder model was trained on normal flight data to learn reconstruction patterns and detect anomalies based on reconstruction error analysis at the 95th percentile threshold. To provide labelled anomalies, we created synthetic abnormal patterns that we labelled to assist in evaluating our model's sensitivity to detecting potentially anomalous data. The results showed a varying achievability of anomaly detection, as well as proof of concept for machine learning based anomaly detection in encrypted UAS telemetry data and potential future use in real-world contexts and edge device applications

Due to the negative environmental impact of international trade through the maritime sector, this industry is gaining disproportionate attention. The development of 'green' maritime technologies is a necessary response to global pressures that seek to reduce greenhouse gas emissions and increase sustainability. Automation through artificial intelligence, including route optimization, autonomous navigation, and smart port systems, is a groundbreaking approach to reducing fuel and emission dependencies. This chapter analyzes the shifts in maritime operations through the lens of the integration of automation and artificial intelligence in 'green' technologies, the regulatory and real-world aspects of compliance, the new environmental benefits, challenges, and opportunities in the quest for 'sustainable' maritime operations.

Artificial Intelligence (AI) is changing how the marine business works in a big way. It is changing how environmental monitoring, security, cargo management, and navigation are done. AI-powered solutions are now used in the marine industry for things like predictive maintenance, route optimization, real-time data analysis, and managing ships on their own. In the past, these activities relied on manual decision-making and analog instruments. These "machine minds at sea" help keep the world's oceans safe, make things run more smoothly, cut down on mistakes made by people, and make things safer. AI also makes it possible to manage the environment in a proactive way by using sustainable routing, pollution monitoring, and emission tracking. Combining human knowledge with AI makes the global maritime ecosystem safer, more efficient, and more sustainable. This chapter talks about how AI is used, what its pros and cons are, and what it means for marine intelligence.

As autonomous vessels and maritime drones transition from experimental technologies to operational realities, they are redefining the structural, regulatory, and ethical foundations of the global maritime industry. While these innovations promise enhanced efficiency, safety, and environmental performance, they also present a range of complex challenges that extend beyond technical feasibility. This chapter provides a holistic, interdisciplinary analysis of the regulatory, ethical, and economic issues of maritime automation. It examines how international frameworks, especially the ones shaped by the International Maritime Organization, are struggling to come to terms with the implications of autonomy. It also identifies some of the most critical gaps in legal accountability, certification, and cybersecurity. Ethically, the chapter questions the transfer of navigational responsibility to machines, the possibility of replacing maritime labor, and dangers of opaque or prejudiced artificial intelligence systems. Economically, it measures the transformative potential and the disruptive effects of automation, especially focusing on capital costs, labor market changes, and new business models. By means of systems-level analysis, the chapter highlights the interdependence of these domains and positions for anticipatory, inclusive governance frameworks. It ends up with concrete recommendations for policymakers, technologists, and industry leaders on how maritime innovation can be technologically sound, ethically grounded, and economically sustainable.

This chapter will consider the transformations brought about in functions performed by military-grade unmanned aerial vehicles (UAVs)-that is, fighter drones-as it concerns their role in contemporary maritime operations and naval warfare. Networked communications, sensor technologies, autonomy, and artificial intelligence (AI) are paving the way in the new opportunities to fill the area over which UAVs will become permanent materials in the strike missions, anti-submarine warfare, search and rescue, and environmental monitoring which can, prior, be only covered by ISR. This chapter also discusses the literature on core theory and current operations practice, the technology behind UAV effects in adverse marine environments, and capabilities and functions with which these technologies are associated.

This chapter per the authors presents the design and evaluation of a smart intrusion detection system tailored for restricted maritime zones, where safeguarding assets against unauthorized access is a pressing challenge. The objective is to integrate motion analysis with spatio-temporal processing to enhance surveillance reliability in dynamic marine environments. The proposed framework employs background subtraction, optical flow techniques, and deep learning models such as CNNs and LSTMs to capture movement patterns over time. A prototype was developed and tested using drone-captured video feeds, demonstrating improved accuracy, reduced false alarms, and efficient detection under varying conditions. Results highlight the system's potential to strengthen maritime security operations by providing real-time situational awareness and supporting regulatory compliance. The chapter concludes that such AI-driven surveillance solutions are essential for advancing autonomous maritime security, with implications for ports, naval facilities, and offshore platforms.

This study investigates the strategic deployment of drones in Taiwan's maritime domain, analyzing their opportunities and operational benefits. The research explores the increasing integration of Unmanned Aerial Vehicles (UAVs) in the maritime sector and their impact on enhancing safety, efficiency, and productivity. Through a needs-based analysis and technology assessment, key application areas such as port monitoring, environmental surveillance, and search and rescue operations are identified. The study also addresses the challenges and barriers to drone implementation, including regulatory considerations, technological limitations, and social-economic factors, proposing solutions and mitigation strategies to facilitate effective drone integration in Taiwan's maritime sector.

Preface

Emerging Technologies in Vessels and Automated Drones for Maritime Innovation is one means of grappling with the wish to comprehend this discontinuity. Blast of the ocean digital and physical technology still the backbone of the globe's commerce, vitally important for the country's security and international environmental management, is what this book targets. Practically, this book will attempt to provide a comprehensive account as well as a discerning scrutiny on the shift to advanced automation and astonishingly high-grade ASV and UAS, along with sophisticated data centric and AI-driven navigational and operational techs onboard these vessels.

The scope of the transformation in the range of construction of vessels, logistics, search and rescue to even ecological monitoring is truly astounding. The last era was characterized by the gradual alteration in the efficiency of the fuel and the construction of the hull. The current era is characterized by a tremendous increase in data and autonomy.

This volume examines the impact of the use of these tools on the world's sustainability initiatives and the consequent improvements in the operational efficiencies, personnel, and cargo safety of Evergreen. Here, the automation of vessels and the accompanying mechanization of decision frameworks, perception strategies, sensor fusion, and vessel autonomy elucidate how intelligence outpaces mechanization. This includes detailed descriptions of remote proactive control and of the planning and collision control modules. These descriptions also include ship-to-shore control integration for remote monitoring and proactive weather analytics systems for routing and optimization.

CHAPTER OVERVIEW

Chapter 1. AI-Enabled Multi-Sensor Drones for Real-Time Maritime Search and Rescue

This chapter presents an innovative framework for enhancing maritime search and rescue operations through AI-enabled, multi-sensor drones. The authors demonstrate how the integration of thermal, RGB, and LiDAR sensors with edge computing enables rapid data processing and real-time decision-making in complex marine environments. By leveraging the AI4SAR dataset and optimizing a YOLOv5s model for efficiency and precision, the study showcases a robust and highly accurate solution for maritime emergencies. The chapter highlights the transformative potential of intelligent UAV systems in improving safety, responsiveness, and reliability at sea.

Chapter 2. Anomaly Detection in MAVLink Telemetry Using Autoencoders

Focusing on the security of UAV communication systems, this chapter explores a machine learning approach to detecting anomalies within MAVLink telemetry data. The authors develop an Autoencoder-based model capable of identifying irregular behavior without relying on predefined attack signatures. By employing encryption methods such as AES and ChaCha20, and executing the model in Python and TensorFlow, the study illustrates how unsupervised learning can enhance UAV resilience against cyber threats. The chapter's findings underscore the potential of anomaly detection systems to safeguard drone operations within sensitive maritime applications.

Chapter 3. Green Maritime Technologies: The Role of Automation in Reducing Emissions and Enhancing Sustainability

This chapter examines how automation is redefining sustainability in maritime operations. With rising global demands for emission reduction, the authors explore how technologies such as autonomous navigation, AI-driven route optimization, and smart port systems are being employed to cut fuel consumption and greenhouse gases. The discussion bridges environmental imperatives with technological innovation, evaluating the benefits, challenges, and regulatory aspects of maritime automation. The chapter underscores the need for collaborative global frameworks to ensure that the maritime sector aligns with the broader goals of ecological responsibility and sustainable development.

Chapter 4. Machine Minds at Sea: A New Era of AI-Driven Maritime Intelligence

The fourth chapter delves into the expanding influence of artificial intelligence across maritime domains, framing AI as the machine mind transforming sea operations. It discusses applications in predictive maintenance, route optimization, and autonomous navigation, emphasizing how AI-driven insights improve efficiency, safety, and environmental monitoring. The author articulates a vision where human expertise and AI coexist, ensuring smarter, safer, and more sustainable ocean management. This chapter captures the essence of the digital revolution at sea and the opportunities it presents for global maritime intelligence systems.

Chapter 5. Regulatory, Ethical, and Economic Considerations

As maritime automation matures, new regulatory and ethical questions arise. This chapter offers an interdisciplinary perspective on the evolving legal and moral dimensions surrounding autonomous vessels and drones. The author examines international frameworks led by the International Maritime Organization, highlighting gaps in accountability, certification, and cybersecurity. Ethical debates around the delegation of navigational authority to machines and the implications for maritime labor are central to the analysis. The chapter concludes by emphasizing the need for governance models that balance innovation with transparency, human safety, and economic fairness.

Chapter 6. Role of Fighter Drones in Maritime Applications

This chapter explores the strategic evolution of fighter drones in naval and maritime contexts. The authors examine how networked communication, AI, and sensor technologies have transformed drones from surveillance tools into key assets for strike missions, anti-submarine warfare, and rescue operations. By reviewing operational theories and technological advancements, the chapter illustrates how UAVs are redefining maritime defense and intelligence. The analysis highlights the growing integration of these aerial systems in modern naval strategy, emphasizing their adaptability and tactical significance in dynamic oceanic environments.

Chapter 7. Smart Intrusion Detection System for Restricted Maritime Zones: Using Motion Analysis and Spatio-Temporal Processing

In this chapter, the authors introduce a smart intrusion detection framework designed to secure restricted maritime zones through AI-powered motion analysis. By combining background subtraction, optical flow, and deep learning architectures such as CNNs and LSTMs, the system achieves improved accuracy and real-time surveillance capabilities. Field testing with drone-captured video validated the model's efficiency in identifying threats and reducing false alarms. The chapter demonstrates how intelligent vision systems can fortify maritime security infrastructures, ensuring better protection for ports, naval facilities, and offshore installations.

Chapter 8. The Strategic Deployment of Drones in Taiwan's Maritime Domain: Opportunities and Operational Benefits

This chapter investigates the growing role of UAVs in Taiwan's maritime operations, focusing on their strategic benefits and implementation challenges. Through a comprehensive assessment of technology and policy, the author identifies key applications in port monitoring, environmental surveillance, and emergency response. The discussion also considers regulatory, economic, and technological barriers, proposing actionable strategies for smoother integration. By positioning drones as essential tools for maritime efficiency and safety, the chapter offers a roadmap for national and regional adaptation in the evolving landscape of maritime technology.

Chapter 9. Waves in the Cloud: Rethinking Maritime Infrastructure in a Connected World

The final chapter reimagines maritime infrastructure in the era of digital connectivity. It presents a compelling analysis of how cloud computing, AI, IoT, and edge technologies are transforming traditional maritime operations into data-centric ecosystems. By enabling real-time coordination, predictive maintenance, and intelligent logistics, these innovations are reshaping ports and vessels into smart entities. The author highlights both opportunities and challenges, ranging from cybersecurity risks to digital inequities, underscoring the importance of governance and collaboration. This chapter concludes the volume by envisioning a resilient, inclusive, and technologically advanced maritime future.

This is a significant aspect. The drone infrastructure also has its UAS. No longer passive surveillance devices, they also perform high-precision hull inspections, map dynamic ice, monitor emissions plume plumes, and compliance during and post emission discharges and constraint proration, and a delivery system augmented to deliver urgent supplies to offshore platforms. The UAS also enhances Vertical Takeoff and Landing Autonomous Surface Vessels (ASVs) to hydrographic surveys and persistent security patrol cover, as well as the collection and oceanographic over long-duration campaigns. It is voiced in: the sensors are pitched against the risk of losing lives bold, brave, statistically.

The UAS optimizes Vertical Takeoff and Landing Autonomous Surface Vessels (ASVs) to hydrographic surveys and persistent security patrol cover, as well as the collection and oceanographic over long-duration campaigns. It is voiced in: the sensors are pitched against the risk of losing lives bold, brave, statistically.

Considerable social upheaval will not happen on its own. Many technical and ethical issues are also highly intricate and must be resolved to ensure security and widespread acceptance. In this book, the authors go beyond merely indicating the potential of the advancements and thoroughly examine the potential pitfalls. Fully integrated smart vessels and cooperative drones will be particularly vulnerable to cyber attacks. The issues of resilient cyber-physical systems, safe remote data transmission, robust, reliable, and completely autonomous operation in remote, open, hostile maritime environments.

Furthermore, the neoclassical international law of the sea, development of which mostly dates back to the 20th century, even finds it greatly challenging to entertain the notion of a vessel operating without a human commander on board. We review the activities of the International Maritime Organization (IMO) regarding the development of new codes for Maritime Autonomous Surface Ships (MASS). This also includes crucial questions of legal liability and insurance, as well as the question of who is legally liable in the event of an incident involving an unmanned asset. The productivity gains greatly shift the functions of sea-going personnel. This book addresses the ethics of departing from a governing posture in crises, especially those that in some respects defy an overriding impossibility, such as the protection of the environment and the recovery of materials from shipwrecked vessels. It further addresses the future of maritime advocacy, which bolts for the gap on re-skilling and up-skilling to bridge the gap of redundant seamanship to remote control operations, analytical frameworks, advanced data systems, and high-level engineering frameworks. The near future maritime professional would not primarily act as a navigator, but rather as a fundamental overseer of the system, and chief data processor and analyst.

This book aims to be a point of reference for all during these novel times. It seeks to assist marine engineers, naval architects, people in the maritime business, fleet owners, scholars and scientists, policymakers and regulators, and all custodians of the science and Mother Nature.

Chapter 1
AI-Enabled Multi-Sensor Drones for Real-Time Maritime Search and Rescue

Jayabhaduri Radhakrishnan
https://orcid.org/0000-0003-3473-1533
Alliance School of Advanced Computing, Alliance University, Bengaluru, India

R. Shekhar
Alliance School of Advanced Computing, Alliance University, Bengaluru, India

Abstract

Maritime Search and rescue (MSAR) solutions must be intelligent, nimble, and autonomous due to the growing frequency and complexity of maritime emergencies, which are caused by climate change, growing coastal populations, and increased marine traffic. This study introduces a multi-sensor, AI-enabled unmanned aerial vehicle (UAV) system intended to improve maritime SAR operations in real time. To provide synchronized data collection, in-situ processing, and quick decision-making, the suggested framework combines thermal, RGB, and LiDAR sensors with a Pixhawk 4 flight controller and a Jetson Xavier NX edge computing module. The domain-specific AI4SAR dataset is used to train YOLOv5s, a compact deep learning architecture that is enhanced with both synthetic and real marine scenes to guarantee resilience in a range of environmental circumstances. Transfer learning, ONNX export, and TensorRT acceleration are used to optimize the model, allowing for inference speeds exceeding 30 frames per second with 92.1% precision, 88.7%

DOI: 10.4018/979-8-3373-3546-9.ch001

recall, and 90.3% mAP@0.5. The outcomes highlight the system's applicability for use in actual SAR situations, where operational resilience and real-time item detection are essential for saving lives in hazardous maritime environments.

I. INTRODUCTION TO MARITIME SEARCH AND RESCUE

Search and Rescue (SAR) operations in maritime environments have found its significance in ensuring the safety and survival of persons during emergencies like capsized boats, vessels going missing, or distress situations. In the Maritime Search-And-Rescue(MSAR) operation, survival is often time-dependent, necessitating rapid response. In recent years, technological developments, specifically UAVs with cameras and sensors, have enhanced the ability to locate people at sea. In the past ten years, deep learning techniques have facilitated the development of automatic systems for identifying people from aerial imagery (Carneiro, 1988); Gozalan et al, 2020; Martinez-Esteso et. al, 2024; Martinez-Esteso et. al, 2025).

Hence, rescue operations at sea to either secure or save and assist people's survival during emergencies like boats capsizing, missing vessels, or distress cases are paramount. But those earlier SAR operations with time constraints and vast expanses of the oceans were hard to carry out. The maritime search and rescue operations have been made difficult because of dynamic conditions, resource constraints and uncertainties (Panić et. al, 2021). The typical hierarchical command structure of rescue operations is not suitable for such situations. The incorporation of artificial intelligence into MSAR will improve decision making, adaptability, autonomy and resource optimization (Kilic et al., 2025; Quero et al, 2025)

That is where drone technology comes in, along with its integration of AI in that technology, giving these SAR missions a paradigm shift (Pettersson, 2020; Ghazlane et. al,2025). AI-assisted drones, equipped with the most sophisticated sensors such as thermal imaging, LiDAR, and high-definition cameras, can now conduct real-time surveillance and object detection, which are the lynchpins of the efficient and precise conduct of maritime SAR operations. Throughout the majority of critical parameters, EUAVDet manages to outperform its competitor methods in terms of FLOPs, mAP, and FPS parameters. EUAVDet-n, the smallest model, with only 1.34 million parameters, provides near-real-time functioning at above 20 FPS on the Jetson Nano. This proves that the model maintains a great balance between detection accuracy and inference speed, inherently favoring the model most suited for deployment on edge-based unmanned aerial vehicle platforms (Wu et. al,2024, Yucesoy et. al,2025).

Increased frequency in severity of maritime emergencies under climate change, rising sea traffic, and poor safety measures demand more agile, intelligent, and reliable SAR solutions. As the coastal population and marine industries grow, so does the complexity of rescue missions that must now contend with several challenges: congested shipping lanes, unpredictable oceanographic conditions, and so on. On the contrary, the limitations of human-led search efforts delayed mobilization, poor visibility at night and with stormy weather, and the very physical scale of oceanic search areas that highlight the actual need of technological enhancement (Bajwa, 2025).

AI-enabled UAV systems represent a convergence of advances in several fields, such as robotics, sensor technology, data science, aerospace engineering, and maritime operations. They are involved not just to replace human effort but rather to complement it, extending the reach, speed, and precision of rescue missions beyond what has ever been possible before. These systems work as intelligent agents, mapping areas autonomously, detecting survivors via thermal and optical imagery, and assessing threats by interpreting real-time environmental inputs. Thus, the importance of swarm intelligence and collaborative drone networks is turning out to a promising horizon in maritime SAR. In large-scale search operations, multiple drones can coordinate with each other via decentralized algorithms to brush up search performance with the adaptability required to cope with the changes happening on the go, such as reassigning search areas based on the newly received data about sudden weather conditions. Such flexibility would allow the search to achieve better coverage and chances to avoid missing target critical in vast search zones (Abi-Zeid, 2019; Hashesh, 2022, Pal, 2024).

Regulatory frameworks are critical in another aspect, that is, defining the application domains of drones and AI pertaining to humanitarian missions (Rejebet. Al, 2021). Such responsible applications include data privacy, airspace regulation, cross-border coordination, and algorithm transparency. In addition, there has been growing acknowledgment from Governments and International Organizations on the strategic contribution that autonomous systems can deliver to disaster response and, subsequently, increased investments in standardization, protocols for testing, and integration to the national rescue infrastructures. Standardization has thus become an emerging impetus for further development in these fields. Therefore, AI-enabled multi-sensor drones are expected to play a crucial role in SAR operations in the near future (Papyan et. al, 2024, Eren et. al, 2025). Indeed, their propensity to provide highly precise sight of the situation in real time-without human risk-being pivotal in the rescuing of maritime casualties makes them really indispensable for the future of emergency response. This chapter focuses on the technical framework, operational plans, and case studies associated with such systems to give a better perspective on how they are making a change in search and rescue dynamics at sea. AI In Drones Algorithms have a life jacket, a survivor, an emergency signal, or a capsized boat being caught in a high-accuracy differentiation. It takes real-time data input for

SAR teams to enhance the speed of decision-making thus creating time savings in deployment (Wang et. al, 2025).

A) Challenges Encountered with Conventional SAR System

The maritime Search and Rescue (SAR) operations should be important. But still, in spite of this importance, traditional methods that are corroborated by extent and feasibility did not get rid of the usual bumps that otherwise would otherwise keep shunning them on their effective lines of operations-helicopters, ships, and manual reconnaissance missions all seem to suffer from many operational, environmental constraints or human considerations, and so there would sometimes be delays that hamper all these approaches in SAR missions (Benz, 2021).

1. Time Sensitivity and Delayed Response

Maritime emergencies unfold rapidly, and the survivability window for individuals stranded at sea is often measured in hours. Conventional SAR units, particularly those located far from the incident site, face logistical delays in deployment. Preparing aircraft, assembling crew, and navigating to the affected location can consume valuable time, reducing the chance of successful rescue.

2. Environmental and Visibility Constraints

Most SAR operations are susceptible to adverse weather incidents such as stormy skies, strong winds, fog, or nightfall, all creating visibility deterioration would consequently put rescue personnel at risk. In many cases, helicopters are relegated to no take-off or landing on safety grounds: on the one hand, ship-based searching becomes ineffective, mainly due to limited sensor range and harsh sea conditions. These scenarios drastically reduce the effectiveness of traditionalism in bad environments.

3. Human Limitations and Fatigue

The human factor has to do with physical endurance, cognitive overload, and situational awareness, all three of which play a major role during extended missions or high-stress situations. Fatigue and an impaired state of concentration can cause missed detections or slower reactions and compromise the efficiency and safety of SAR operations.

4. Insufficient Area Coverage

The ocean being quite an expanse, its dynamic environment makes it a gigantic challenge for any kind of operations to achieve complete coverage. Ships and helicopters have a restricted distance they can cover in a particular time, and manual searchers frequently leave gaps or else are redundant. Sea currents are immensely unpredictable, thereby making it even more difficult to project the direction of survivor drift.

5. High Operating Costs

Traditional search-and-rescue methods are resource-consuming in terms of fuel, maintenance, and the cost of personnel. The permanent availability of seeking helicopters or patrol boats for search and rescue is a big monetary concern. Thus, it is often limiting them from being made available for constant or large operations due to lack of funds.

6. Limited Data Analysis and Use

Real-time data analysis for effective choices-making is something that does not exist in the conventional system. Although monitoring tools like radar, sonar, and visual checks are in effect, the absence of integrated data analytics means insights from environmental sensors or previous missions are often undervalued or inaccessible throughout a crisis.

The Chapter is organized as follows: Chapter 1 highlights the critical role of Maritime Search and Rescue (MSAR) in saving lives at sea, addressing key challenges such as limited response times, resource constraints, and communication inefficiencies. Chapter 2 explores how Artificial Intelligence (AI) can enhance MSAR operations, improving decision-making, resource optimization, and coordination through machine learning, computer vision, and autonomous systems. Chapter 3 presents the architecture of an AI-powered drone system equipped with multiple sensors for maritime SAR along with tasks and functionalities. Chapter 4 discusses about the implementation of the proposed AI-enabled drone system for maritime SAR operations. This chapter also summarizes the development and deployment of the AI-enabled drone system for maritime search and rescue operations. Chapter 5 discusses about conclusion and future work.

II. ROLE OF AI AND UAVS IN MODERN SAR OPERATIONS

Deep learning is a disruptive technology for maritime Search and Rescue, providing solid solutions for complex tasks such as object detection, classification, localization, and situational awareness in dynamic marine environments. CNNs have been prominently utilized in the detection of survivors, vessels, and floating objects in aerial and satellite images, particularly because of their spatial hierarchy-learning capability. RNNs along with Long Short-Term Memory (LSTM) networks are employed to analyse movement patterns over time as the sequential data. Newer methods based on Transformers and attention mechanisms direct computational resources toward unconcerned parts of images, thus aiding the detection of small or partly occluded objects (Awais et al, 2025; Tjia et al, 2024). Once further tailored and optimized for embedded deployment (such as NVIDIA Jetson or ARM-based platforms), these deep learning models promise high-accuracy and low-latency performance of real-time, autonomous SAR systems.

For small objects in the image or objects that are partially buried in the background, a Transformer-based approach aimed to improve recall when applied to analyze SAR images (Perera et al., 2022; Youk et. al, 2023). Advanced models including YOLOv5/YOLOv8, Faster R-CNN, EfficientDet, and RetinaNet have been modified for real-time detection on edge devices such as drones and maritime robots. Hybrid methods fusing RGB, thermal, and LiDAR inputs through multi-modal fusion network have boosted detection confidence under low visibility, nighttime, and heavy weather conditions. In addition, GANs have been applied to synthetic data generation to mitigate data availability issues on account of the rarity of the SAR scenarios (Jin et. al, 2025).

GGT-YOLO is a cutting-edge object identification algorithm specially designed for maritime drone cruising. It modifies the traditional feature extraction process by creating Transformer modules to broaden its detection of smaller or occluded objects, often found in open water scenarios. The study also mentioned that it balances detection accuracy with computation efficiency so that the algorithm works for any UAV with resource constraints, according to (Li et al., 2022).

Drones play a crucial role in remote sensing and smart surveillance; however, their limited onboard processing power affects object detection performance. To bridge the gap, researchers propose a new algorithm that strikes a balance between detection precision and computational cost for drone-based object detection in large-scale maritime applications. Transformer enhances feature extraction, as a result, detection of small or occluded objects, is improved, while convolutions favoured for detection are replaced with simpler linear transformations to save computation. Using a drone image dataset from maritime settings, our method has increased precision, recall, and average precision, with a reduction in parameters by 11.6%

and FLOPs by 7.3%, enhancing the feasibility of drone usage in various maritime and remote sensing applications (Zhang et al., 2023).

Detecting small objects in images and videos continues to create a key challenge in intelligent transportation and autonomous systems; such objects include pedestrians, small vehicles, or maritime obstacles. These objects usually occupy very small pixel spaces given as $\leq 32 \times 32$ and therefore limit the viewable context information required in applying accurate detection. Traditional object detection methods fail at this challenge. To overcome it, researchers have turned their attention toward deep learning and computer vision approaches that are targeted at small object detection SOD (Rekavandi et. al, 2025).

Lightweight, deep learning-based ship detection network named YOLOv5-ODConvNeXt, is built on YOLOv5 by integrating Omni-dimensional Convolution (ODConv) to the model for enhanced accuracy with no increase in model size, whereas ConvNeXt replaces the original C3 block of YOLOv5, to enhance speed with minimal loss in accuracy. Evaluation on a custom drone-view dataset with 3,200 images yielded 48.0% detection accuracy, 1.2% better than that of YOLOv5s, while the inference time was reduced to 8.3 ms per image, 13.3% faster than YOLOv5s (Cheng S et al., 2023; Zaman et al, 2023).

For the success of drone-based search and rescue operations in inefficient weather or visibility conditions, it is necessary to have a robust person detection system under aerial imagery. But the postures of the people are various, human typologies are sparse, they appear small due to viewing from a very high angle, and they may have poor visibility. All of these parameters combined will make feature extraction very difficult and detection accuracy low. (Zhang et al.,2025).

Researchers present the modular plug-and-play optimization strategy using the YOLOv8 framework and optimizes the performance in real time without sacrificing accuracy. Using the SeaDronesSee dataset, the proposed technique increased accuracy by 13.53% reduced computational cost by 85.63% and achieved over 20 times the detection faster than the official two-stage detector of speed while achieving similar accuracy (Zhao B et al.,2024).

III. ARCHITECTURE OF AI-ENABLED SAR DRONES FOR MARITIME SAR OPERATIONS

The architecture of an AI-enabled multi-sensor drone system for maritime SAR operations, composed of several key components, which work collaboratively to ensure smooth and efficient search and rescue missions, are as follows: UAV Platform, Sensor Suite Module, Flight Planning and Control Module, Data Acquisition and

Synchronization Module, AI Model Training and Inference Module, Performance Evaluation Module and Deployment Optimization Module as depicted in Figure 1.

Figure 1. Module Components In AI-Enabled Drones For Maritime SAR Operations

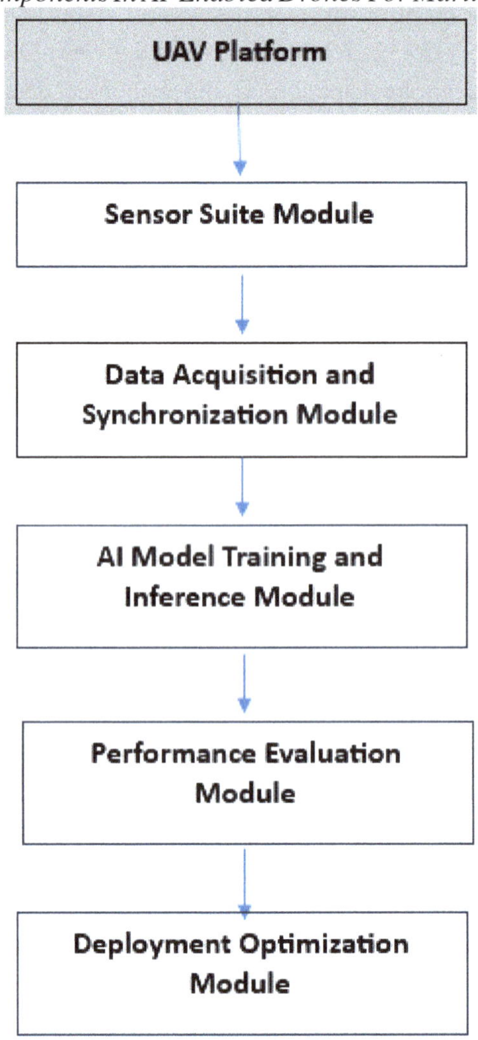

1. UAV Platform Module

The heart of an artificial intelligence component within a multi-sensor array is an unmanned aerial vehicle (UAV), thus serving as the basic platform for air deployment in maritime SAR operations. Multi-rotor or fixed-wing UAVs are used in either case. Both types have their own advantages depending on the mission type: multi-rotor UAVs can maneuver very well to achieve vertical take-off and landing; hence, they remain suitable for hovering while surveying a confined or turbulent space in the marine environment. Fixed-wing UAVs, on the other hand, offer endurance with long-distance surveys and reconnaissance work, especially if the missions go on for hours across long stretches of ocean. Whatever type, all these UAVs are designed to endure complex yet harsh weather phenomena, with ruggedized airframes protected with components capable of withstanding the hostile marine environment comprising variable winds, limited exposure to saltwater, and varying weather. These conditions alone create robust platforms for UAVs where intelligent search-and-rescue systems are built. They support uninterrupted aerial vigilance and data acquisition under adverse operational environments.

2. Sensor Suite Module

The Sensor Suite Module is the fundamental building block defining the AI-enabled UAV system by incorporating a variety of calibrated sensors necessary for extremely robust maritime situational awareness. One of these sensors is the FLIR Duo Pro R thermal imaging camera responsible for capturing infrared signatures to detect heat-emitting objects, such as human survivors, even under reduced visibility conditions, including fog or blacked-out operations. In parallel, a Sony IMX378 RGB camera provides high-resolution visible data for real-time object detection and classification tasks. For three-dimensional spatial analysis, a Velodyne VLP-16 LiDAR sensor is used to derive high-resolution 3D point clouds to help accurately determine object profiles and environmental features such as debris or vessel structures. Moreover, Global Positioning System integrated with Inertial Measurement Unit sensors, allows accurate geospatial localization along with tracking of orientation, are critical to the operation of autonomous flight control, sensor fusion, and data geo-referencing during SAR missions.

3. Data Acquisition and Synchronization Module

In the UAV framework, the viewpoint of synchronous multi-source data fusion and accurate post-processing mainly depends on time synchronization. All sensor measurements, including those captured by a thermal and RGB sensor, by LiDAR, GPS, or IMU, are precisely time stamped either by Network Time Protocol (NTP) or by either of the two methods: MAVLink-based synchronization, allowing an integrated temporal framework that can serve to synchronize and align the asynchronous streams from the sensors. This temporal alignment is vital for the collection of sensor data for effective data fusion purposes and reliable object detection for fast operation in maritime settings.

The system additionally enables the use of Robot Operating System (ROS) tools, particularly rosbag, to record the synchronized data streams for a deeper data-gathering exercise. These logs constitute a rich multi-modal dataset, which contains thermal imaging, visual information, 3D point clouds, position, orientation, and all accurately time-stamped measurements. The support for real-time processing and mission reliability is given to the UAV by NVIDIA Jetson Xavier NX as its onboard edge computing platform, which processes AI models locally and furthermore serves as a local data storage unit. Depending on mission requirements and communication availability, the collected data is stored either locally or transferred via RF (Radio Frequency) or LTE communication links to a ground control station for remote monitoring, further analysis, or real-time decision support.

4. AI Model Training and Inference Module

YOLOv5s is the object detection model in use, which is selected for its fast inference and good accuracy, fit for real-time edge implementation on the NVIDIA Jetson Xavier NX. Its small architecture provides for rapid and almost real-time detections of SAR-relevant targets such as victims and debris in maritime situations. The training was performed on the AI4SAR dataset with more than 5,000 labeled maritime frames with RGB, thermal, and LiDAR modalities. The classes included Survivors, Life Vessels, and Capsized vessels. The fine-tuning of the model was performed by means of transfer learning on a PyTorch-based HP Z4 G4 GPU workstation.

Heavy data augmentation techniques have been used for improving generalization: rotation, noise injection, contrast adjustment, etc., which considered real-world variability. For classification, Binary Cross Entropy is applied during model training and Complete IoU (CIoU) for bounding box regression for classification, with the aim of achieving precise object detection and localization respectively.

5. Performance Evaluation Module

Using the YOLOv5s object detection model, evaluation was carried out on a separate test set containing 1,000 labeled maritime SAR frames. The model exhibited a Precision of 92.1%, Recall of 88.7%, and mean Average Precision (mAP@0.5) of 90.3%, thereby confirming very high detection accuracy. Concerning deployment efficiency, real-time inference is obtained with the model at 28 ms/frame, which is well above 30 FPS on the NVIDIA Jetson Xavier NX, making it suitable for onboard edge processing during time-critical SAR missions.

6. Deployment Optimization Module

The model is initially exported in the Open Neural Network Exchange (ONNX) format to accommodate different platforms and deployments on various hardware environments. Then it is further enhanced through model optimization with NVIDIA TensorRT using FP16 quantization and model compression techniques. Applying these optimizations reduces the model size by 38%, making it more memory-device efficient. Furthermore, its inference speed is improved by 20%, which is critical for quicker execution on real-time search and rescue operations. This also minimizes the required memory footprint for running the optimized model, enabling concurrent applications, such as object and navigation detection, without performance deterioration in the overall system.

Key Tasks and Functionalities of AI-Enabled UAVs in Maritime Search and Rescue Operations

The tasks performed by the AI-enabled drones during maritime search and rescue operations are as follows:

1. Autonomous Navigation and Flight Control

The UAVs operate using an intelligent flight control system that can perform missions autonomously. The autonomous flight control system would enable the UAVs to be programmed and carry out advancement missions independently. This translates into pre-programmed route planning, which allows flying drones according to specific objectives of the mission along pre-defined flight paths. Also, where dynamic path alterations come into play, the UAV can adjust its route during the real-time phase according to environmental changes and obstacles along its flight or target identification. This adaptive ability helps in less collision and faster response to changing situations. All these are among the best features that can be incorporated into an advanced mobile model applicable in the efficient and reliable exercise of maritime SAR operations.

2. AI-Driven Decision-Making

Artificial Intelligence driven intelligent detection along with predictive analysis capabilities are the core cognitive elements of the drone system relevant for a maritime search-and-rescue mission. Famous object detection algorithms including YOLO and Faster-RCNN take visual and thermal data as input and accurately identify and classify objects of interest such as life vests, human figures, capsized vessels, or an SOS marker. In conjunction, the movement prediction model uses the environmental parameters such as ocean currents, winds, and weathering patterns to predict the probable drift trajectories of the survivors or objects and thus, limit the focus of searching. Sensor fusion techniques merge data from several sensors to reduce false positives while improving detection accuracy and overall situational awareness in support of quicker and sounder decision-making during SAR operations.

3. Real-Time Communication and Control Systems

A robust communication infrastructure is needed to ensure uninterrupted and useful data transmission between UAV and SAR command units during maritime search and rescue operations. Therefore, data links providing satellite, RF, or cellular connectivity are needed to relay streams of validated video, real-time sensor outputs, and AI-generated assessments to operators on the ground. To reduce latency and thereby increase responsiveness, edge computing units carried by the drone locally process sensor data for immediate detection and classification of possible targets while not relying solely on remote servers. HMIs with user-friendly dashboards consolidate all pertinent information for mission operators: live telemetry, visual feeds, geospatial coordinates, and urgent alerts—to help operators make decisions quickly and knowledgeably in dynamic SAR situations.

4. Mission Coordination and Integration

These AI-enabled drone systems are generally incorporated under the large umbrella of search and rescue (SAR) systems to mold effective coordination and an efficient SAR operation. This means that interoperability with maritime vessels and helicopters and coast bases play an important role in multi-modality response coordination, thus ensuring a unified and holistic approach towards SAR operations. Post-operation evaluations are aided through cloud-based mission logging and analytics, permitting mission performance analysis down to the deepest detail. In this way, data facilitates pattern recognition when dealing with sometimes occurrence incidents; this information is also beneficial for improvement of the AI model features and operational tactics, further enhancing future SAR missions' effectiveness.

IV. IMPLEMENTATION AND RESULTS

This chapter outlines the experimental setup, datasets used, evaluation metrics, and key results, demonstrating improvements in detection accuracy, response time, and overall system efficiency.

A) Drone System Setup

The AI-enabled drone system used in maritime search and rescue is a very sophisticated platform, meant to easily integrate with advanced sensing, navigation, and processing technologies. The quadrotor design enables outstanding maneuverability to operate in confined and challenging environments that are typically endured during a maritime rescue mission. The thermal imaging camera detects heat signatures from survivors or distressed vessels even in conditions inaccessible to the naked eye, such as low visibility or at night, which is critical for effectiveness in SAR operatives.

Velodyne VLP-16, the 16-channel LiDAR sensor, one of the major components of detailed 3D environment mapping, enables the drone to have the real-time capability of detecting and avoiding obstacles while accurately locating survivors or objects in the water. The high-definition RGB camera (Sony IMX378) affords an opportunity for the drone to capture high-resolution well-suited for both SAR operators and AI algorithms in real time. It essentially ends up being the most critical part of recording an event that could be more useful in assessing and taking decisions after the event has taken place. The Pixhawk 4-based flight control system forms the backbone of UAV autonomous capabilities. This is complemented by GPS and Inertial Measurement Unit modules, allowing accurate assurance of drone position and orientation in exceedingly adverse weather conditions or very turbulent ocean.

B) Curated Drone Dataset

For data acquisition using an AI-enabled drone placed for maritime SAR operations, the systematic process is followed that encompasses calibration of sensors, planning flight, logging data, and synchronization. The drone operates with a Pixhawk 4 flight controller integrated with GPS and IMU modules and is equipped with a FLIR Duo Pro R thermal camera, a Velodyne VLP-16 16-channel LiDAR sensor, and a high-resolution Sony IMX378 RGB camera. Before the deployment, all sensors are calibrated: LiDAR calibration is carried out using ROS-based tools for ensuring accurate 3D point-cloud generation, and thermal and RGB cameras are synchronized to a common timestamp framework, most frequently through the use of MAVLink or NTP protocols.

The flight paths are pre-planned by mission planning software such as QGroundControl or Mission Planner, where GPS waypoints, altitude settings, and image overlap ratio settings are defined to ensure the best coverage of the maritime area set as target. During the flight, the sensors continuously collect data, the thermal camera takes infrared imagery with GPS metadata, the RGB camera records high-resolution images, and the LiDAR streams point clouds that are logged with ROS tools like rosbag. These data streams are either stored locally on the onboard Jetson Xavier NX or transmitted via RF or LTE links to a ground control station. After the mission, the collected data—comprising synchronized RGB, thermal, LiDAR, GPS, and IMU logs—is fused to provide a rich dataset. This multi-modal dataset can then be annotated using tools like CVAT or LabelImg to prepare training data for object detection models, thereby enabling robust AI-based maritime SAR operations.

C) Model Training and Evaluation

To facilitate real-time intelligent object detection in maritime search and rescue operations, aerial data-visual, thermal, depth-was utilized in training deep learning models with potential for high-speed, highly accurate inference in edge computing environments. The onboard processing platform selected for deployment is the NVIDIA Jetson Xavier NX, an extremely compact and highly energy-efficient AI edge device whose performance is 21 TOPS (Tera Operations Per Second)-efficient. This makes it suitable for running deep learning models on UAVs in critical missions.

D) Model Selection: YOLOv5s for Edge Deployment

The object detection model selected for the system is YOLOv5s, a compact version of the YOLO (You Only Look Once) family that carries the brightest futures when it comes to real-time detection enabled by an extremely optimized architecture. YOLOv5s stands out as a compromise among speed, accuracy, and computational efficiency, making it especially ideal for edge AI scenarios, where low latency and minimized energy consumption are required. It uses Cross Stage Partial and Path Aggregation Network for efficient feature aggregation and enhancement in multi-scale detection, which is valuable in a maritime environment due to the size variation of objects as far as distance and perspective are concerned.

E) Model Training

Training has been accomplished on several states using selected datasets named AI4SAR with more than 5,000 labeled image frames from missions carried out by drones with synchronized RGB/visible cameras, thermal infrared sensors, and LiDAR. For every image frame, bounding boxes and class labels were annotated according to objects of SAR-related importance. They include Survivors (human figures, either floating or partially submerged), Life vests, Capsized or overturned vessels and Emergency signals or flares.

This curated dataset is augmented with both synthetic and real imagery to ensure robustness and generalizability concerning the very diverse real-world maritime conditions such as illumination levels, wave intensities, or weather patterns. The process provides generative modeling of synthetic images via simulations of maritime scenes, allowing control over variation in objects' pose, illumination, and occlusion. To improve variability within the resulting training samples, data augmentation techniques include the rotation of the imagery, noise injection, contrast enhancement, and horizontal flipping, among others.

For faster computation, the model is trained using PyTorch on a high-performance HP Z4 G4 GPU workstation, with a batch size of 16 over 200 epochs with the help of Adam and a learning rate scheduler for smooth, adaptive learning. This model uses Binary Cross-Entropy (BCE) to work on the classification tasks and Complete Intersection over Union (CIoU) loss for bounding box regression tasks. All input images are resized and normalized to 640×640 pixels, as final input to maintain homogeneity and also to optimize the detection performance. To accelerate convergence, transfer learning is employed to initialize the network with pre-trained YOLOv5s weights, originally trained on the COCO dataset, before fine-tuning it to domain specific AI4SAR dataset. This strategy enables the model to train itself adequately and thus makes it useful in the maritime SAR-specific object detection tasks.

F) Model Evaluation and Performance Metrics

After successful training, the model has been evaluated on a held-out test set comprising 1,000 frames collected from separate drone missions that were not included in the training phase. Thus, it guarantees the unbiased assessment of the model's generalization ability. The evaluation carried out using standardized object detection performance metrics, giving a precision of 92.1%, a recall of 88.7%, and Mean Average Precision (mAP@0.5) of 90.3%, demonstrates the model's ability to detect and accurately localize important objects in difficult maritime settings, is depicted in Figure 2.

Figure 2. Model Performance Metrics on SAR Test Set

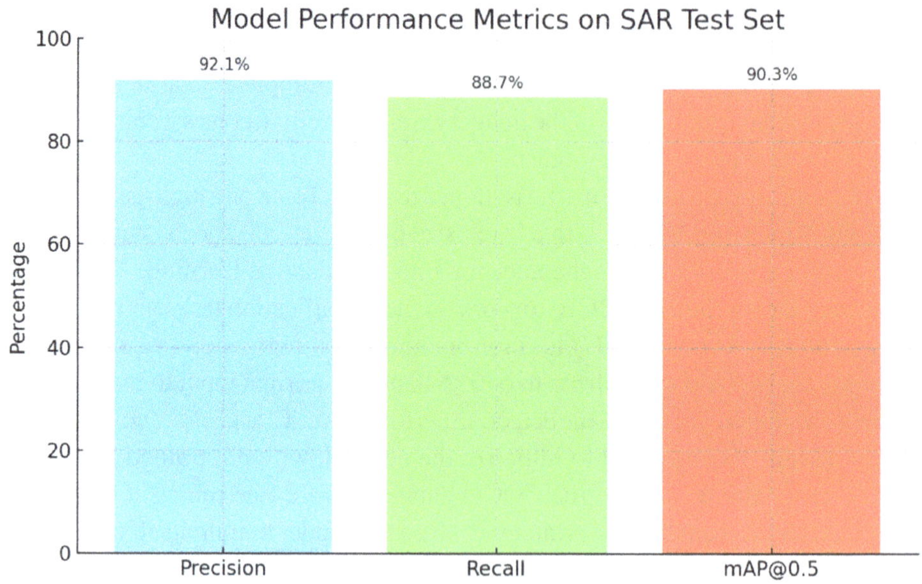

Also, the optimized model has a low inference time of 28 ms/frame on the NVIDIA Jetson Xavier NX, which allows it to work at above 30 frames per second (FPS). These results prove that the model is not only highly accurate but also quick to respond in terms of real-time ability, thus making it a strong candidate for assured deployment in SAR-type drones in tough environmental conditions.

G) Deployment Optimization

The model has to first export the ONNX (Open NeuralNetwork Exchange) format to improve its flexibility in deployment on different platforms and ease its process of deployment. Then NVIDIA TensorRT, which performs intelligent optimization techniques—such as precision calibration (FP16) and model compression—was used to fine-tune the model. The optimizations reduce the model's size by 38%, which helps it become lightweight and more manageable on edge devices. This gain in size comes simultaneously with a 20% speed boost in inference, so the drone can do data processing and decision-making at greater speed, a key requirement for real-time engagement. Due to all these changes, the model will also take up less memory, allowing for the concurrent execution of other tasks by the drone, such as object detection and navigation, without issues. All these improvements are built to ensure that the system can tackle the intricate and time-bound search-and-rescue missions without wasting resources—ideal for real deployment scenarios where reliability and efficacy are paramount.

Table 1 summarizes the model performance before and after optimization using ONNX export and NVIDIA TensorRT. From the results shown in Table 1, it is evident that the model optimization performed using ONNX export and NVIDIA TensorRT, significantly enhances the efficiency of the AI model, deployed on Jetson Xavier NX edge device. Reduction in model size by 38% makes the model more lightweight and suited for memory-constrained environments. Inference Speed is improved by 20%, thereby enabling quick decision-making. Increase in Frame Rate from 28 FPS to 36 FPS, ensures real-time performance critical for time-sensitive maritime SAR operations.

Table 1. Summary of the Model Performance Before and After Optimization using ONNX export and NVIDIA TensorRT

Metric	Before Optimization	After Optimization	Improvement
Model Format	PyTorch (.pt)	ONNX + TensorRT (FP16)	Flexible deployment
Model Size	52 MB	32.2 MB	↓ 38%
Inference Speed	35 ms/frame (~28 FPS)	28 ms/frame (~36 FPS)	↑ 20% faster
Memory Footprint	High	Reduced	Enables multitasking
Deployment Platform	Limited (x86-based)	Edge (Jetson Xavier NX)	Broader compatibility
Task Parallelism	Limited	Supported	Detection + Navigation

The model performance before and after optimization, showing improvements in model size, inference speed, and frame rate (FPS) portrayed in Figure 3, concludes that these optimizations enable the UAV to perform object detection and navigation concurrently with high responsiveness, making it ideal for real-time, resource-constrained SAR scenarios.

Figure 3. Model Performance Before and After Optimization

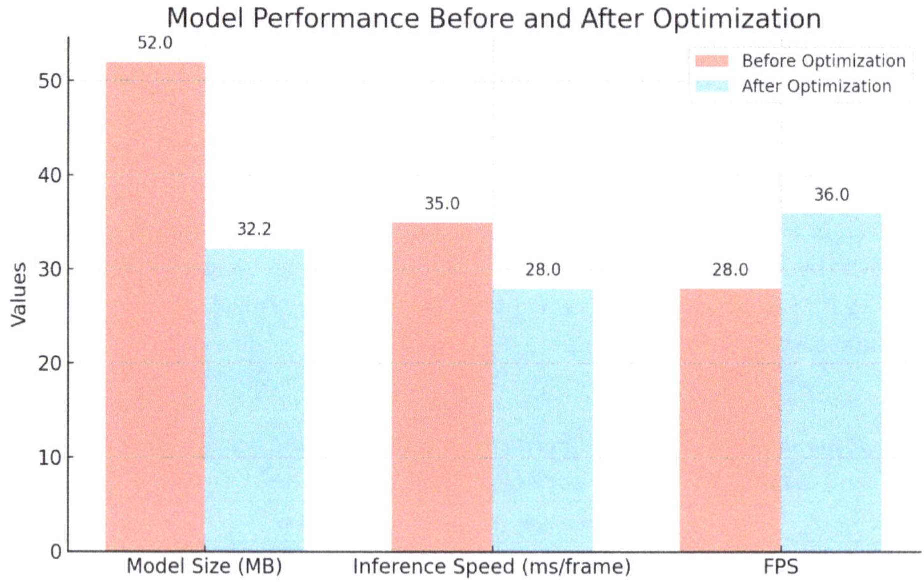

H) Field Validation

In the field trials conducted for the system, controlled maritime environment trials simulated possible situations that would be encountered by a drone during search and rescue missions. The field trials entailed drones with RGB cameras, thermal and LiDAR sensors attached to them, and thus permitted modeling of how well the model works with different types of environments on the water. A very good performance was shown for the system tested among its series of comparative tests in difficult situations, such as changing sea states, illumination (day to dusk), and weather conditions. It also showed good performance in detecting important objects such as survivors, life vests, capsized boats, and emergency signals in unfavorable conditions. For example, the model could still identify many objects, such as during the transition from day to dusk or in situations with poor lighting, due to sensor fusion-the fusion of thermal and RGB data making the model more superior in this respect. The greatest benefit of using both thermal and RGB data simultaneously became obvious when it proved possible to detect survivors trapped or concealed behind waves or thermal noise. Sometimes, thermal alone would not be enough because the objects are submerged or occluded by water, but when combined with RGB data, it gave the model a better chance in detecting partially hidden survivors, thus raising significantly its recall-hence, reducing the chance of it missing any important objects, however challenging the conditions.

The trials proved that the system performed really well in the low-visibility conditions that pertain to fog and slight rain. In those cases, the synergy of thermal and RGB sensors allowed the drone to see objects like a capsized boat partially submerged in the water or a life vest when those objects were too difficult to see using one type of sensor alone. The fusion of both sensor data ensured that the drone did not miss any important object, thus enhancing its reliability. The onboard system, utilizing the NVIDIA Jetson Xavier NX, performed just fine during the testing. It allowed the drone to do processing and make real-time changes to its flight path depending on what it detected. All of this capability for real-time decision-making is important for SAR missions where seconds really matter.

V. CONCLUSION AND FUTURE WORK

This research work describes a full AI-based drone system aimed at providing a useful improvement in maritime search and rescue operations. Deployment of lightweight object detection models, multi-sensor data fusion, and edge computing via the NVIDIA Jetson Xavier NX provided the system with truly impressive real-time capabilities. Field validation showed that the model could successfully detect survivors, life vests, capsized vessels, and emergency signals under widely varying environmental situations, with high precision, recall, and mAP. Transfer learning, ONNX exportation, and TensorRT optimization facilitated low-latency inference and seamless onboard processing, with timely decisions that are critical in SAR scenarios.

Improving contextual understanding during rescue missions can further enhance this system by adding environmental parameters such as wave patterns, wind speed, and tide levels. Expanding the AI4SAR dataset with more varied real-world video footage, including nighttime or stormy scenarios, would increase model robustness. Ongoing research will also focus on the autonomy of coordination between multiple drones via swarm intelligence and real-time communication. Partnership with SAR authorities for integration with their communication protocols will also be pursued to ensure the practical deployment and scaling of the system in real-world scenarios.

REFERENCES

Abi-Zeid, I., Morin, M., & Nilo, O. (2019, May). Decision Support for Planning Maritime Search and Rescue Operations in Canada. In *ICEIS (1)* (pp. 328-339). DOI: 10.5220/0007730303280339

Awais, C. M., Reggiannini, M., Moroni, D., & Salerno, E. (2025). A Survey on SAR ship classification using Deep Learning. *arXiv preprint arXiv:2503.11906*.

Bajwa, A. (2025). AI-based emergency response systems: A systematic literature review on smart infrastructure safety. *Available at SSRN* 5171521.

Benz, L., Münch, C., & Hartmann, E. (2021). Development of a search and rescue framework for maritime freight shipping in the Arctic. *Transportation Research Part A, Policy and Practice*, *152*, 54–69. DOI: 10.1016/j.tra.2021.07.006

Carneiro, J. P. (1988). Maritime Search and Rescue. *IETE Technical Review*, *5*(3), 111–114. DOI: 10.1080/02564602.1988.11438248

Cheng, S., Zhu, Y., & Wu, S. (2023). Deep learning based efficient ship detection from drone-captured images for maritime surveillance. *Ocean Engineering*, *285*, 115440. DOI: 10.1016/j.oceaneng.2023.115440

Eren, V., & Duman, H. (2025). Artificial Intelligence Support In Disaster Management. *Kamu Yönetimi ve Teknoloji Dergisi*, *7*(1), 13–36. DOI: 10.58307/kaytek.1580460

Ghazlane, Y., Hajar, B., Hajar, E. M., Saida, H., & Mourad, E. Y. (2025). Artificial Intelligence in Drone and Anti-drone Systems: Cutting-Edge Advances and Applications. In *International Conference on Electrical Systems and Smart Technologies* (pp. 167-174). Springer, Cham. DOI: 10.1007/978-3-031-86705-7_16

Gozalan, A., John, O., Lubcke, T., Maier, A., Reimann, M., Richter, J. G., & Zverev, I. (2020). Assisting maritime search and rescue (SAR) personnel with AI-based speech recognition and smart direction finding. *Journal of Marine Science and Engineering*, *8*(10), 818. DOI: 10.3390/jmse8100818

Hashesh, A. O., Hashima, S., Zaki, R. M., Fouda, M. M., Hatano, K., & Eldien, A. S. T. (2022). AI-enabled UAV communications: Challenges and future directions. *IEEE Access : Practical Innovations, Open Solutions*, *10*, 92048–92066. DOI: 10.1109/ACCESS.2022.3202956

Jin, Z., He, T., Qiao, L., Duan, J., Shi, X., Yan, B., & Guo, C. (2025). MES-YOLO: An efficient lightweight maritime search and rescue object detection algorithm with improved feature fusion pyramid network. *Journal of Visual Communication and Image Representation, 109*, 104453. DOI: 10.1016/j.jvcir.2025.104453

Kilic, K. I., Maity, S., Sung, I., & Nielsen, P. (2025). Challenges and AI-driven solutions in maritime search and rescue planning: A comprehensive literature review. *Marine Policy, 178*, 106692. DOI: 10.1016/j.marpol.2025.106692

Li, Y., Yuan, H., Wang, Y., & Xiao, C. (2022). GGT-YOLO: A novel object detection algorithm for drone-based maritime cruising. *Drones (Basel), 6*(11), 335. DOI: 10.3390/drones6110335

Martinez-Esteso, J. P., Castellanos, F. J., Calvo-Zaragoza, J., & Gallego, A. J. (2024). Maritime Search and Rescue Missions with Aerial Images: A Survey. *arXiv preprint arXiv:2411.07649.*

Martinez-Esteso, J. P., Castellanos, F. J., Calvo-Zaragoza, J., & Gallego, A. J. (2025). Maritime search and rescue missions with aerial images: A survey. *Computer Science Review, 57*, 100736. DOI: 10.1016/j.cosrev.2025.100736

Pal, O. K., Shovon, M. D., Mridha, M. F., & Shin, J. (2024). In-depth review of AI-enabled unmanned aerial vehicles: Trends, vision, and challenges. *Discover Artificial Intelligence, 4*(1), 1–24. DOI: 10.1007/s44163-024-00209-1

Panić, I., Ćelić, J., Bistrović, M., & Škrobonja, A. (2021). Drone as a part of maritime search and rescue operations. *Technologies, Techniques and Applications Across PNT*, 63.

Papyan, N., Kulhandjian, M., Kulhandjian, H., & Aslanyan, L. (2024). Ai-based drone assisted human rescue in disaster environments: Challenges and opportunities. *Pattern Recognition and Image Analysis, 34*(1), 169–186. DOI: 10.1134/S1054661824010152

Perera, M. V., Bandara, W. G. C., Valanarasu, J. M. J., & Patel, V. M. (2022, July). Transformer-based SAR image despeckling. In *IGARSS 2022-2022 IEEE International Geoscience and Remote Sensing Symposium* (pp. 751-754). IEEE. DOI: 10.1109/IGARSS46834.2022.9884596

Pettersson, E. (2020). Design of a drone system for maritime search and rescue missions.

Quero, C. O., & Martinez-Carranza, J. (2025). Unmanned aerial systems in search and rescue: A global perspective on current challenges and future applications. *International Journal of Disaster Risk Reduction, 118*, 105199. DOI: 10.1016/j.ijdrr.2025.105199

Rejeb, A., Rejeb, K., Simske, S., & Treiblmaier, H. (2021). Humanitarian drones: A review and research agenda. *Internet of Things : Engineering Cyber Physical Human Systems, 16*, 100434. DOI: 10.1016/j.iot.2021.100434

Rekavandi, A. M., Xu, L., Boussaid, F., Seghouane, A. K., Hoefs, S., & Bennamoun, M. (2025). A Guide to Image-and Video-Based Small Object Detection Using Deep Learning: Case Study of Maritime Surveillance. *IEEE Transactions on Intelligent Transportation Systems, 26*(3), 2851–2879. DOI: 10.1109/TITS.2025.3530678

Tjia, M., Kim, A., Wijaya, E. W., Tefara, H., & Zhu, K. (2024). *Enhancing Robustness of Human Detection Algorithms in Maritime SAR through Augmented Aerial Images to Simulate Weather Conditions.* arXiv preprint arXiv:2408.13766.

Wang, Y., & Han, Y. (2025). Measuring maritime search and rescue (SAR) accessibility using an improved spatiotemporal two-step floating catchment area method: A case study in the South China Sea. *Journal of Transport Geography, 124*, 104155. DOI: 10.1016/j.jtrangeo.2025.104155

Wu, W., Liu, A., Hu, J., Mo, Y., Xiang, S., Duan, P., & Liang, Q. (2024). EUAVDet: An efficient and lightweight object detector for UAV aerial images with an edge-based computing platform. *Drones (Basel), 8*(6), 261. DOI: 10.3390/drones8060261

Youk, G., & Kim, M. (2023). Transformer-based synthetic-to-measured SAR image translation via learning of representational features. *IEEE Transactions on Geoscience and Remote Sensing, 61*, 1–18. DOI: 10.1109/TGRS.2023.3267480

Yucesoy, E., Balcik, B., & Coban, E. (2025). The role of drones in disaster response: A literature review of operations research applications. *International Transactions in Operational Research, 32*(2), 545–589. DOI: 10.1111/itor.13484

Zaman, F. H. K., Tahir, N. M., Yusoff, Y. M., Thamrin, N. M., & Hasmi, A. H. (2023). Human detection from drone using you only look once (YOLOv5) for search and rescue operation. *Journal of Advanced Research in Applied Sciences and Engineering Technology, 30*(3), 222–235.

Zhang, X., Feng, Y., Wang, N., Lu, G., & Mei, S. (2025). Aerial Person Detection for Search an d Rescue: Survey and Benchmarks. *Journal of Remote Sensing, 2025*, 0474.

Zhang, Y., Yin, Y., & Shao, Z. (2023). An enhanced target detection algorithm for maritime search and rescue based on aerial images. *Remote Sensing (Basel)*, *15*(19), 4818. DOI: 10.3390/rs15194818

Zhao, B., Zhou, Y., Song, R., Yu, L., Zhang, X., & Liu, J. (2024). Modular YOLOv8 optimization for real-time UAV maritime rescue object detection. *Scientific Reports*, *14*(1), 24492. DOI: 10.1038/s41598-024-75807-1

Chapter 2
Anomaly Detection in MAVLink Telemetry Using Autoencoders

Mahira Mohamed Mowjoon
University of Technology Sydney, Australia

Emi Sekikawa
University of Technology Sydney, Australia

Shuaimin Zhang
University of Technology Sydney, Australia

ABSTRACT

In this chapter, we describe an anomaly detection system for UAS telemetry data based on unsupervised Autoencoders. We created telemetry transmissions with encrypted communications using AES and ChaCha20 in Python, following the MAVLink protocol, to maintain a realistic communications environment. The encrypted logs were transformed to CSV files and pre-processed with normalization and feature selection to prepare the data for model training. An Autoencoder model was trained on normal flight data to learn reconstruction patterns and detect anomalies based on reconstruction error analysis at the 95th percentile threshold. To provide labelled anomalies, we created synthetic abnormal patterns that we labelled to assist in evaluating our model's sensitivity to detecting potentially anomalous data. The results showed a varying achievability of anomaly detection, as well as proof of concept for machine learning based anomaly detection in encrypted UAS telemetry data and potential future use in real-world contexts and edge device applications

DOI: 10.4018/979-8-3373-3546-9.ch002

1. INTRODUCTION

Unmanned aerial vehicles (UAVs), more commonly referred to as drones, have seen an uptick in adopting commercial, agricultural, and research environments in recent years. Increasingly, these drones are becoming part of an Internet of Things (IoT) ecosystem, enabling a variety of telemetry data to be wirelessly communicated to cloud services or ground control stations. Such direct integration improves the capabilities for autonomous operations but opens additional ways for cybersecurity threats to occur. UAVs often operate under significant hardware constraints such as low computational power and battery capacity and are often designed to communicate via other insecure protocols such as the MAVLink or HTTP protocols (Ficco et al., 2022).

MAVLink (Micro Air Vehicle Link) is a widely known communication protocol used for UAVs which is designed as a lightweight, open-source messaging system that allows for telemetry to be communicated over TCP or UDP. The MAVLink protocol transmits data in plaintext, with 2.0 version only providing optional signature-based authentication, and no complete encryption of the contents of the messages (Allouch et al., 2019). Though vulnerable when it encounters any kind of CRC or checksum error, attacks on the data streams can be classified as spoofing attacks, message tampering attacks, replay attacks, or modified command attacks. In addition to those threats, UAV systems expose several risks and vulnerabilities closely related to UAV-specific threats as GPS spoofing, sensor data spoofing, sensor data manipulation, and abnormal command input in UAV command systems are risks that only manifest in autonomous, battery-limited, and mobile threat aspects (Yahuza et al., 2021).

Conventional intrusion detection systems (IDS), like Snort or Suricata, generally depend on detection based on rule-based detection and on known attack signatures; however, these techniques are not relevant for UAVs because real-time behaviour is erratic, and attacks are usually new. To address this gap, our project applied **Autoencoder-based anomaly detection** (Yao & Zheng, 2023), a form of unsupervised machine learning that learns to reconstruct normal operational patterns from unlabelled data and flags anomalies based on reconstruction error (Chollet, 2018).

In this work, we implemented a complete simulation framework for detecting anomalous behaviour in UAV telemetry streams using Autoencoders. We used Python to simulate MAVLink message traffic and generated synthetic telemetry data streams that included GPS coordinates, altitude, roll/pitch/yaw values, and sensor readings such as temperature. Since not publicly available MAVLink-based telemetry datasets were suitable for cybersecurity evaluation, we generated realistic MAVLink-style data and manually embedded a variety of anomalies for model training and validation.

The design principles on which our anomaly detection system was built were attained successfully as follows:

- Efficient Deployment: The Autoencoder architecture was designed to perform optimally under limited computational resources, making it well-suited for deployment on resource-constrained UAV hardware, such as flight control units or edge processors. The model only employs basic layers to maintain low memory and computational demands while still maintaining performance levels (Chollet, 2018; Arafat & Mohsin, 2022).
- Unsupervised Learning: Because there is no labelled dataset for UAV telemetry attacks, we decided to use an unsupervised method. Autoencoder was trained only with normal MAVLink telemetry data, thereby learning normal operational behaviour latent representations. This enables the model to detect anomalies as deviations from learned behaviour, without any attack signatures ((Yao & Zheng, 2023; Kumar et al., 2021).
- Clarifiable Output: For interpretability, the model provides visualizations of reconstruction error measures for each input sample. A key feature that allows a cyber analyst or system integrator to understand the cause of any deviation, such as whether it was due to injected noise, falsified coordinates, or command changes (Zheng et al., 2020; Yao & Zheng, 2023).
- Protocol-Level Envelope: Through end-to-end simulated MAVLink messages from ground control station to onboard flight computer, including encryption and transmit logic. The model will be tested to conditions that are similar to real UAV operations, allowing for contextual and efficient testing (Allouch et al., 2019; Ficco et al., 2022).

Samanth et al. (2022) defines three key components of the UAV system that we simulated: IoT drone, flight computer, and ground control station (controller). The IoT drone is the onboard component that receives, processes, and acts upon control and telemetry messages. The flight computer is the primary mechanism on the actual drone that performs navigation logic processing, sensor management, and data encryption or decryption. The controller exists as a ground control station that creates MAVLink messages, uses encryption processes, and sends the messages in UDP packets to initiate communications with the drone. Together, they simulate a realistic case of drone communications with onboard telemetry control.

2. LITERATURE REVIEW

As drones continue to proliferate in a variety of sectors, concerns regarding cybersecurity are becoming more visible. According to researchers (Yahuza et al, 2021), drones are particularly vulnerable to communication-phase attacks such as GPS faux signals, data injections, message modification, and malware-based attacks.

Regarding communication protocols, some researchers (Allouch et al., 2019) point out that while the code (communication protocol) successfully called MAVLink is used most frequently, it lacks any security features and is especially vulnerable to replay and modification attacks. Some researchers have built protocols, called MAVSec, which enhance authentication and encryption.

Some researchers have shown that ChaCha20 is superior to traditional algorithms, especially in resource-constrained environments like drone systems (Allouch et al., 2019). However, other researchers (Cecchinato et al, 2023) have shown more recently that AES encryption works extremely well in many conditions at the application layer, which is useful in terms of securing UAV communications. They have shown that AES ensures real-time encryption and decryption of very high-quality multimedia data, both operations occur with very little latency, ensuring confidentiality and resilience to unreliable radio environments. The per-packet encryption approach, in addition, ensures recoverability in the event of partial data loss, and passive attack simulations showed that without the correct key, the encrypted content remains secure.

A 2024 comprehensive study on the MAVLink protocol (Hamza et al., 2024) established that ChaCha20 is considered one of the fastest symmetric encryption protocols available, preserving the confidentiality of communications while getting good memory usage and CPU efficiency that does not negatively impact overall system performance. Thus, ChaCha20 is well-suited for resource-constrained environments such as embedded systems or unmanned aerial vehicles, UAVs. Alongside ChaCha20, the Advanced Encryption Standard (AES) provides another crucial role in secure communications. Several studies report that a significant contribution to the confidentiality and integrity of data transmission between MAVLink communication is gained by applying AES in Counter (CTR) mode for message encryption, then hashing the key with SHA-256, and then applying the Diffie-Hellman protocol for key exchange. Both encryptions possess two separate advantages; ChaCha20 demonstrates superiority in performance and efficiency, while AES emphasizes message integrity and prevention against tampering. Ultimately, the authors choose both ChaCha20 and AES as the two encryption protocols to secure MAVLink messages, providing a balance between system efficiency and secure communication.

Another important avenue of research is in detecting anomalous behaviour in UAV telemetry without supervision through deep learning methods such as Autoencoders. Traditional rule-based IDSs generally do not detect unknown attacks or anomalous behaviour, which is often the case in dynamic UAV scenarios. Recent research (Chollet, 2018; Yao & Zheng, 2023) shows that Autoencoders may learn normal UAV operational patterns with unlabelled data and suggest anomalous behaviour based on reconstruction error. The Autoencoder is a type of neural network that learns to project input data to a hidden representation and back. When trained on recordings if normal UAV telemetry, such as altitude or velocity, the model learns to reconstruct a normal flight trajectory with low error. If an anomalous payload is injected into the model, such as spoofed GPS coordinates or sensor values, it will create a high reconstruction error and suggests an anomaly.

In addition, Abdalla et al. (2022) proposed a new framework for anomaly detection within UAVs involving stacked autoencoders and a temporal filtering module. The combined model improved the accuracy of anomaly detection in the telemetry of drones for various flight patterns by modelling spatial and temporal dependencies. The results of their experiments indicate a significant improvement performance over baseline autoencoder architectures and a benefit in the reduction of false alarms for non-stationary UAV flight operations.

In addition to the Autoencoder-based approaches, Sun et al. (2020) demonstrated a Gated Recurrent Units-based deep learning anomaly detection technique in drone telemetry. The study utilized temporal dependencies in the flight data while the level of detection performance improved with subtle anomaly detection over time. This study suggests that augmenting temporal context into machine learning techniques will improve UAV cybersecurity.

Unlike supervised models, Autoencoders do not require labelled attack data and thus can be utilized to detect novel or zero-day attacks in MAVLink-based drone systems. Furthermore, their lightweight architecture is also helpful for deploying to edge devices, i.e., drone-mounted processors and for facilitating real-time anomaly detection in resource-limited environments.

This study utilizes the previous literature by integrating protocol-level MAVLink simulation and Autoencoder-based anomaly detection into a single experimental process with an interpretable, scalable, and protocol-aware approach to UAV cybersecurity.

3. IMPLEMENTATION AND EVALUATION

This chapter presents the methodological framework followed in the execution of the anomaly detection prototype. The entire process, data ingestion, preprocessing, model development, training, and testing, was conducted with Python in conjunction with Google Colab.

Google Colab was selected as the development platform since it provides free GPU and TPU access, real-time collaboration support through integration with Google Drive, and can even eliminate the need for local setup, which works in favour of collaboration and cloud accessibility (Bisong, 2019).It also provides an easy way to run and visualize code in a Jupyter Notebook, so you can experiment in an interactive way. Python was selected to be the primary programming language because of its many possible machine learning libraries (such as Pandas for manipulating data and TensorFlow/Keras for modelling) and preprocessing and assessment libraries (such as Scikit-learn). These libraries are effective because they have a concise syntax, high-level abstractions, and good community support, which are more practical for rapid prototyping and exploratory research (Pedregosa et al., 2011; Chollet, 2018).

In our project, we used the Crypto and pymavlink libraries, where:

- Crypto is used to provide data encryption and decryption service. We used the Crypto.Cipher and Crypto. Random modules to encrypt and decrypt simulated MAVLink messages to add security to communication. Specifically, we used symmetric encryption algorithms such as AES and ChaCha20 to encrypt message payloads and a secure random number generator (such as get_random_bytes) to generate encryption keys and initialization vectors (IVs) to secure the message payloads during transmission and protect it from replay attacks.
- pymavlink is used to parse and create MAVLink protocol packets which allowed us to simulate common UAV telemetry messages like HEARTBEAT and GLOBAL_POSITION_INT.

3.1 MAVLink Simulation

Currently, mainstream IoT drone communication architectures typically involve a ground control station (GCS) acting as a bridge between the drone and the Internet. The GCS exchanges messages with the drone via radio links, while the drone receives these messages through a communication module and forwards them to the onboard computer and flight controller for processing.

Figure 1. IoT Drone Architecture with MAVLink-Based Communication Flow

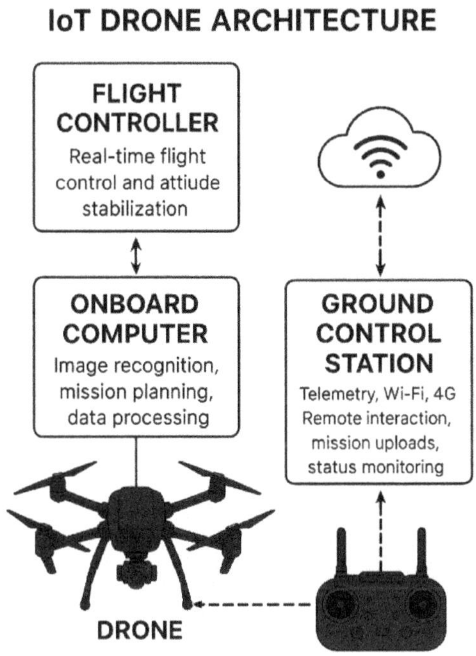

Therefore, mainstream IoT drone architectures typically consist of a flight controller, an onboard computer, and a ground control station or remote controller. The flight controller is responsible for real-time flight control and attitude stabilisation. The onboard computer, usually running a Linux-based system, handles high-level tasks such as image recognition, mission planning, and data processing. The ground control station communicates with the drone via wireless links such as telemetry, Wi-Fi, or 4G, enabling remote interaction, mission uploads, status monitoring, and control command transmission.

3.1.1 The reason why we choose MAVLink:

Several application-layer communication protocols are used in drone systems, including HTTP/HTTPS, MQTT, and MAVLink. For our study, we compared HTTPS and MAVLink.

- HTTPS: A protocol often employed for messaging over the Internet, HTTPS facilitates drones' connection to servers and cloud services. HTTPS uses asymmetric to agree on a symmetric session key, and relies on TLS/SSL and certificate trust chains for authentication. HTTPS is highly secure, but incurs significant overhead and requires TCP, which makes it unsuitable for low-latency control communication. Additionally, the majority of security research regarding HTTPS is focused on Internet communication more generally, rather than UAV specific applications.
- MAVLink: In contrast, the MAVLink is a lightweight, open-source protocol intended for UAV - GCS communication (Allouch et al., 2019). MAVLink has an efficient encoding scheme, which makes it useful in low-bandwidth, real time telemetry applications. Regarding experiments, MAVLink has Python implementations that support simulation and parsing of messages, achieving the level of controllability we decided to use.

pymavlink can be leveraged to parse and construct MAVLink protocol packets, allowing simulation of standard UAV telemetry messages, including HEARTBEAT and GLOBAL_POSITION_INT.

Table 1. Comparison of HTTPS and MAVLink

	HTTPS	**MAVLink**
Design Purpose	Secure internet communication for general web communication	Lightweight protocol specifically designed for drone-GCS communication
Typical Use in Drones	Accessing internet services, uploading media, retrieving mission data	Real-time telemetry, control commands between drone and ground station
Encryption	TLS/SSL(built-in)	external security (e.g., via VPN, AES)
Efficiency	Heavier overhead due to encryption and HTTP headers	Highly efficient; designed for low-latency, low-bandwidth environments

The distinction between HTTPS and MAVLink extends beyond merely their protocol capabilities. HTTPS automatically keeps content of the messages hidden due to its built-in encryption, therefore anomaly detection could only use metadata features (e.g., packet timing, frequency, or size distribution). Conversely, MAVLink has telemetry in plaintext unless additions are made for security. While this greatly increases the system's vulnerability, it also allows for richer behavioral features for anomaly detection (e.g., deviations in command sequences, abnormal parameter values). So, to give the anomaly detection task realism in terms of UAV security weaknesses, and technical sophistication in assessing behaviors for anomaly detections, the MAVLink protocol is established as the primary protocol in this research.

3.1.2 The comparison between AES and ChaCha20

Due to the absence of secure encryption mechanisms in MAVLink, we examined the relevant literature to analyze encryption algorithms and arrive at the decision to implement AES and ChaCha20.

AES: A mature and widely-adopted standard that has support for 128-, 192-, and 256-bit keys. It is regarded as highly secure and is the default mode of encryption for protocols such as HTTPS.

ChaCha20: Efficiently designed for devices with limited computational resources, making it an appropriate choice for IoT drones with constrained on-board hardware. It is also immune to timing and cache-based side-channel attacks, since it does not rely on patterns of memory access.

Table 2. Comparison of AES and ChaCha20

Encryption Algorithm	**AES**	**ChaCha20**
Maturity	High	middle
Dependence on hardware devices	High	no
Widespread adoption	High	Middle

The selection of the encryption algorithm has a direct effect on the anomaly detection performance in UAV systems. AES is secure and used in standard settings but typically requires hardware acceleration for the best speed. This is potentially more problematic for smaller resource-constrained drones, where it could introduce latency and, consequently, delay the ability to detect anomalies in real-time telemetry. ChaCha20, on the other hand, is much lighter and works well in software alone, making it more compatible with the types of anomaly detection systems that require a real-time response. In addition, ChaCha20 provides better protection against side-channel attacks, so the encryption layer itself does not become a new source of anomalies or provide a new attack surface. In sum, the comparison between AES and ChaCha20 reveals a basic trade-off: while AES is mature and usable across multiple applications, ChaCha20 enable lightweight, low-latency anomaly detection within the UAV realm.

3.1.3 Simulation of Mavlink Message

Our project focuses on autoencoder-based anomaly detection on MAVLink messages. Once we picked AES and ChaCha20 as our encryption algorithms, we began simulating MAVLink messages. MAVLink is an open-source communication protocol and it supports many message types. Besides differing in sequence numbers,

MAVLink messages are primarily distinguished by the values of their specific fields. We used the pymavlink library to simulate these messages for our experiments.

We simulated different types of MAVLINK messages, such as heartbeat messages and attitude messages. Attitude messages provide the current orientation and angular velocity of the drone in real time. It is essential for navigation, stabilization, and user feedback. Heartbeat messages are sent periodically by the drone to advertise its presence, status, and system type, which is one of the most important messages in any MAVLink connection.

Figure 2. Code of MAVlink Message simulation

```
from Crypto.Cipher import AES
from Crypto.Util.Padding import pad, unpad
from Crypto.Random import get_random_bytes

AES_KEY = b'this_is_my_super_secret_key_1234'  # 32bytes = AES-256
BLOCK_SIZE = 16

def encrypt_data(plaintext: bytes) -> bytes:  3 用法
    iv = get_random_bytes(16)
    cipher = AES.new(AES_KEY, AES.MODE_CBC, iv)
    ciphertext = cipher.encrypt(pad(plaintext, BLOCK_SIZE))
    print("AES method:",iv + ciphertext)
    return iv + ciphertext

def decrypt_data(ciphered: bytes) -> bytes:  2 用法
    iv = ciphered[:16]
    ciphertext = ciphered[16:]
    cipher = AES.new(AES_KEY, AES.MODE_CBC, iv)
    return unpad(cipher.decrypt(ciphertext), BLOCK_SIZE)
```

When simulating the HEARTBEAT message, the values of the fields type, autopilot, base_mode, custom_mode, and system_status were randomized in order to introduce variety to the simulation. In MAVLINK, the different values of these fields signifies a different state of the drone. This type of message in MAVLINK protocol primarily is used by the ground control station (GCS) to identify and connect with the drone, which is why it contains a number of different fields.

The Type field in MAVLink HEARTBEAT message serves as an indicator for the type of unmanned system, thus allowing the ground control station (GCS) to understand what type of vehicle it is communicating with. This is important for the GCS to be able to identify and effectively regain control of a specific type of drone. The Type field uses the enumeration values defined within MAVLink. For instance, type = 2 represents a quadrotor which is one of the most popular types of multirotor drones. Similarly, type = 4 represents an octorotor being used in industrial applications due to its ability to carry a greater payload compared to a quadrotor.

This Type value is useful in helping the system automatically select the appropriate type of control interface and flight logic.

The Autopilot element indicates the autopilot system built within the drone. If the Autopilot value is 4, this means the drone is running PX4 firmware.

The variable base_mode displays the active mode and when it has a value of 81, it indicates that the drone is in manual mode State. The variable custom_mode expands on the details of the flight mode and it is specific to the autopilot mode. The variable system_status gives an overall readiness status of the drone and if it has a value of 4, it indicates the drone is powered on and ready to fly. The variable mavlink_version refers to the version of the protocol being used and it would typically be 3 for MAVLink 2.0. All of these fields in the message will allow the GCS to effectively understand the status and operation mode of the drone.

We also simulated sending ATTITUDE messages by taking a randomized attitude and angular velocity of the drone. ATTITUDE messages are very important for both the GCS and onboard autopilot system to understand the drone's position in space, and rotational speeds about each axis.

In the MAVLink ATTITUDE message, there are several important fields that represent the drone's angular position and velocity in space. This data is critical for both the GCS and the onboard systems to understand and visualize the position and motion of the drone in real time proprioceptively. The arguments and range of values for those fields in the message we simulated are described below:

- **time_boot_ms:** Represents the time in milliseconds since the system was powered on. It is commonly used for synchronization. **Range in simulation**: 0 to 0xFFFFFFFF (unsigned 32-bit). *Example:* 150000 means the drone has been running for 150 seconds.
- **roll:** Indicates the drone's rotation around the **X-axis** (left/right tilt). **Range in simulation**: -3.14 to 3.14 radians (approx. -180° to 180°). *Example:* roll = 0.5 means a slight tilt to the right.
- **pitch:** Describes rotation around the **Y-axis** (forward/backward tilt). **Range**: -1.57 to 1.57 radians (approx. -90° to 90°). *Example:* pitch = -0.3 indicates a slight nose-down tilt.
- **yaw:** Measures rotation around the **Z-axis**, controlling heading/direction. **Range**: -3.14 to 3.14 radians. *Example:* yaw = 1.57 corresponds to 90°, or eastward heading.
- **rollspeed:** Angular velocity around the **X-axis**, in radians per second. **Range**: -0.5 to 0.5 rad/s. *Example:* rollspeed = 0.2 means a slow rightward rolling motion.

- **pitchspeed:** Angular velocity around the **Y-axis**, in radians per second. **Range**: -0.5 to 0.5 rad/s.
  Example: pitchspeed = -0.1 indicates the drone is pitching downward.
- **yawspeed:** Angular velocity around the **Z-axis**, also in radians per second. **Range**: -0.5 to 0.5 rad/s.
  Example: yawspeed = 0.3 suggests a moderate turn to the right.

By randomizing these values within realistic ranges, the simulation creates a dataset of drone flight states for training and testing anomaly detection models.

3.1.4 How to Implement These Two Algorithms Using Python

In this section, we will show you how to implement two symmetric encryption methods, AES and ChaCha20, in Python. Both of these methods are used to secure MAVLink messages sent between the ground control station and the drone. The implementation of the encryption and decryption processes is detailed enough to focus on data confidentiality and data integrity, which are vital in drone communication systems to avoid tampering or an interception. Using cryptographic libraries in Python, we are employing realistic schemes for encrypting and decrypting data in a way that matches the low computational weight and real-time requirements of drone communication.

(A) AES Encryption

Figure 3. Code of AES implantation by Python

```
if message_type == "heartbeat":
    msg = mavlink2.MAVLink_heartbeat_message(
        type=random.randint( a: 0,  b: 30),
        autopilot=random.randint( a: 0,  b: 20),
        base_mode=random.randint( a: 0,  b: 255),
        custom_mode=random.randint( a: 0,  b: 1000),
        system_status=random.randint( a: 0,  b: 10),
        mavlink_version=3
    )

elif message_type == "attitude":
    msg = mavlink2.MAVLink_attitude_message(
        time_boot_ms=int(time.time()*1000) % 0xFFFFFFFF,
        roll=random.uniform(-3.14,  b: 3.14),
        pitch=random.uniform(-1.57,  b: 1.57),
        yaw=random.uniform(-3.14,  b: 3.14),
        rollspeed=random.uniform(-0.5,  b: 0.5),
        pitchspeed=random.uniform(-0.5,  b: 0.5),
        yawspeed=random.uniform(-0.5,  b: 0.5)
    )
```

When we integrated AES encryption, we opted to use CBC (cipher block chaining) mode from the pycryptodome library for AES-256 encryption and decryption. First, we declared a key of length 32 bytes in order to use AES-256 encryption, and specified a block size of length 16 bytes. In the encrypt_data function, each of our encrypting operations generates a random 16-byte initialization vector (IV) to ensure that there's randomness. The plaintext is padded to fit with AES block size, and then encrypted with the AES cipher in CBC mode. The output is the IV plus the ciphertext since both are needed for successful decryption. The decrypt data function, as the name suggests, gets the IV from the front of the data provided, re-creates the ciphertext with the same key and IV, and then decrypts that ciphertext. Finally, we remove the padding to return the original plaintext. This implementation of AES ensures the security and correctness of AES encryption, and these two methods will encrypt and decrypt the MAVLINK signals we want to simulate, as any mechanism for protecting sensitive data in drone communications will have.

The diagram in Figure 4 depicts the visual flow of the AES encryption and decryption processes, as well as the merging of these processes into the MAVLink message transmission. These flows serve to clarify the process of implementing AES and to show how data is securely processed before transmitting and after receiving.

The left diagram illustrates the internal encryption and decryption logic for AES depending on using CBC mode. When encrypting, it is assumed that the input plaintext is first padded to be divisible by the AES block size by using a random 16 byte initialization vector (IV). The IV is then used to create an AES cipher that uses the key and IV. The padded plaintext is then encrypted. For decryption, the IV is extracted as the first 16 bytes, reconstruct the AES cipher and then the remaining ciphertext is decrypted and unpadded to give access to the original plaintext.

The right diagram demonstrates the merging of this encryption mechanism with the MAVLink message. When the MAVLink message is started, it is passed into the encryption process where the key and the IV are generated and the plaintext message is converted to ciphertext. For the receiving side, ciphertext message is received followed by the decryption process using the same shared key and IV and the original MAVLink message is recovered. This establishes new secured reliable message exchange for drone communication.

Figure 4. Visual Flow of AES Encryption and Decryption with MAVLink Message Integration

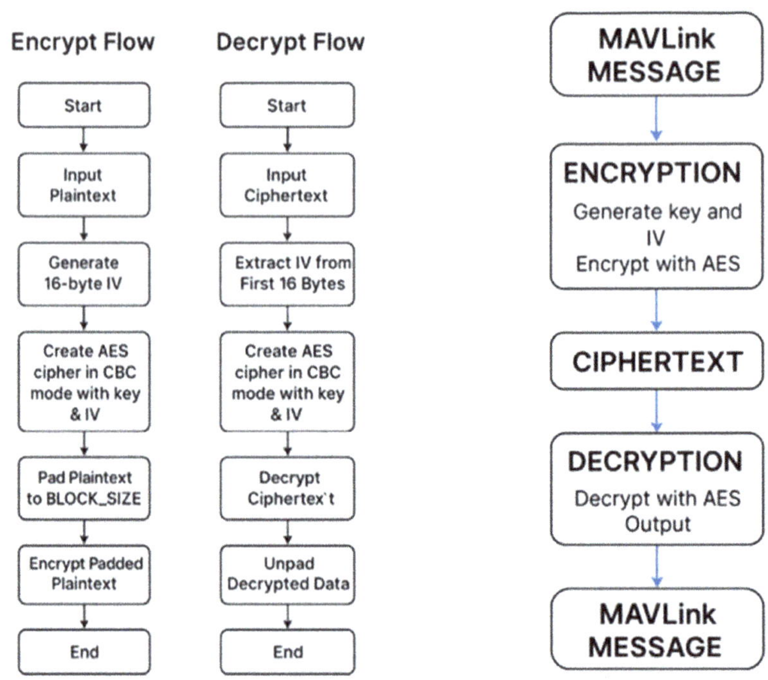

(B) ChaCha20 encryption

Figure 5. Code of ChaCha 20 implantation by python

```
from Crypto.Cipher import ChaCha20
from Crypto.Random import get_random_bytes


CHACHA_KEY = b'supersecretkeyforchacha20test!!!'  # 32 bytes

def encrypt_data_chacha(plaintext: bytes) -> bytes:  2用法
    nonce = get_random_bytes(8)  # ChaCha20 need 8 bytes nonce
    cipher = ChaCha20.new(key=CHACHA_KEY, nonce=nonce)
    ciphertext = cipher.encrypt(plaintext)
    return nonce + ciphertext  # nonce

def decrypt_data_chacha(ciphertext: bytes) -> bytes:  2用法
    nonce = ciphertext[:8]
    encrypted = ciphertext[8:]
    cipher = ChaCha20.new(key=CHACHA_KEY, nonce=nonce)
    return cipher.decrypt(encrypted)
```

Similarly, the pycryptodome library is used to perform ChaCha20 encryption and decryption. To begin ChaCha20, we will define a key of 32 bytes (256 bits) and will generate an 8 byte nonce for each encryption using a random function to provide uniqueness and security. The encrypt_data_chacha will create a cipher object that is created around the key and nonce. It will then encrypt the plaintext and return the concatenation of the nonce and ciphertext. The decrypt_data_chacha will take the nonce back out of the beginning of the input and create the cipher the same way with the key and nonce and decrypt the ciphertext back to plaintext.

3.1.5 Simulation of MAVLink message exchange process

To simulate the MAVLink message exchange process and demonstrate how encryption algorithms can enhance communication security, after randomly generating drone signals and encrypting them using AES and ChaCha20 encryption protocols, we designed a program to completely simulate the process of generating MAVLINK messages, encrypting MAVLINK messages, decrypting MAVLINK messages, and receiving and storing MAVLink messages.

Figure 6. The workflow of MAVLink Communication Python code

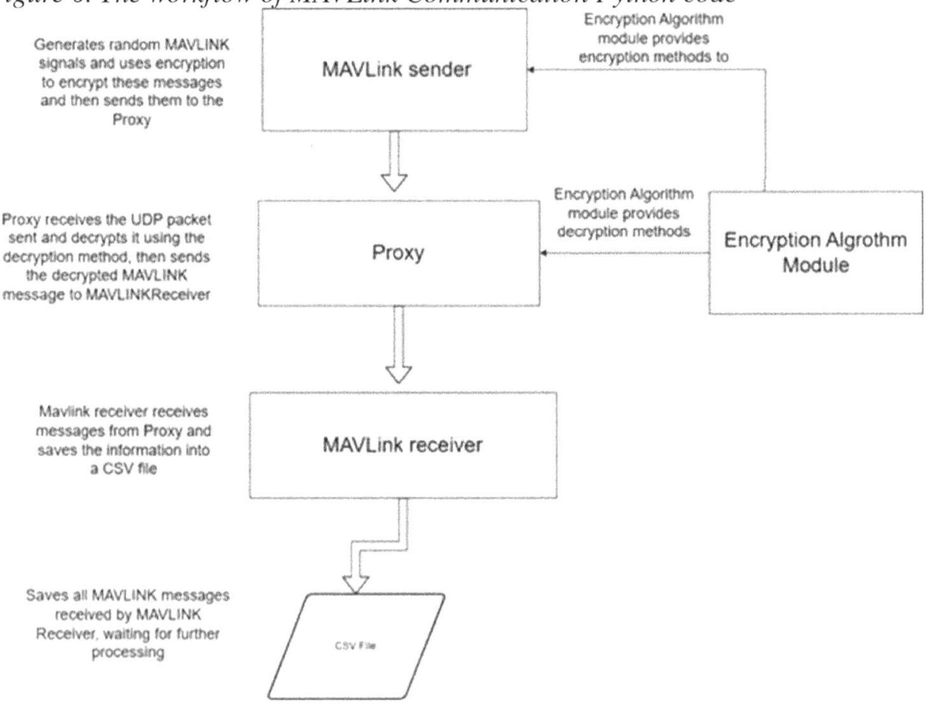

The MAVLink sender generates MAVLink messages and chooses an encryption algorithm to encrypt them.

The message proxy acts like the onboard computer: It gets UDP packets, decrypts the message, and forwards it to the MAVLink receiver. In our project, two proxies have been created. One is called DroneProxy.py, which decrypts MAVLink messages that have been encrypted with AES. The other is called CHACHAProxy.py, which decrypts MAVLink messages that have been encrypted with ChaCha20.

In addition to the MAVLink sender and the message proxy, the Encryption Algorithm Module is an essential part of the MAVLink communication workflow. On the MAVLink Sender side, it provides two different encryption methods (encrypt_data() for AES-CBC and encrypt_data_chacha() for ChaCha20) for the MAVLink messages that have been generated and need to be encrypted for transmission. On the Proxy side, it provides corresponding decryption methods (decrypt_data() and decrypt_data_chacha()) that can decrypt the received cipher text and reconstitute the original MAVLink message before forwarding it to the MAVLink Receiver. This module allows the use of different encryption algorithms in the MAVLink communication workflow. The flexibility of switching between encryption schemes makes it easy to send secure and verifiable messages.

The MAVLink receiver, which simulates the CPU of the drone, receives the MAVLink messages after they have been decrypted, acts on them, and logs them in a CSV file.

In our framework, MavlinkSender uses the Pymavlink library to create MAVLink messages, which are subsequently packed in binary format.

The binary messages are then encrypted using their specific encryption module, e.g., AES, and sent via a UDP socket to the Proxy to be forwarded and processed by the Mavlink Receiver.

Throughout our project, we specified whether to encrypt MAVLink messages using AES or ChaCha20 in the main method of MavlinkSender (shown in Figure 7).

Figure 7. The choice of MAVLink Sender

```
if __name__ == "__main__":
    choice = input("choose chacha20 or AES: ").strip().lower()
    if choice == "chacha20":
        encrypt_method = encrypt_data_chacha
    elif choice == "aes":
        encrypt_method = encrypt_data
    else:
        print("❌ Invalid choice. Defaulting to AES.")
        encrypt_method = encrypt_data

    main()
```

To simulate the MAVLink message communication from a drone, we will run the MAVLinkSender.py, DroneProxy.py, and MAVLinkReceiver.py. When MAVLink-Sender is filed, it will send encrypted messages to the DroneProxy. The DroneProxy will handle the messages, decrypt them, and then send them, now in plaintext form, to the MAVLinkReceiver. This is the entire process of a MAVLink message being transmitted, from the creation of a MAVLink message in the MAVLinkSender, to the successful receipt of the message in the MAVLinkReceiver. During this path (where the message is handled by DroneProxy.py), the program will call the decrypt_data function from AES.py, or decrypt_data_chacha from ChachaProxy.py.

So, this means, to simulate MAVLink communication that is encrypted with AES, I will run the MAVLinkSender.py, DroneProxy.py, and MAVLinkReceiver.py together, ensuring that the full simulation process (i.e. MAVLink message creation, receipt, and decryption) occurs. Once I start the MAVLinkSender program, it will ask me if I want to use one of the two types of encryption - AES or ChaCha20. Once "AES" is selected as the encryptor, the MAVLinkSender.py will create random MAVLink messages and encrypt them using the encrypt_data function from AES.py.

The output in Figure 8 is captured from the console window of MAVLinkSender.py:

Figure 8. Encrypted MAVLink message using AES.

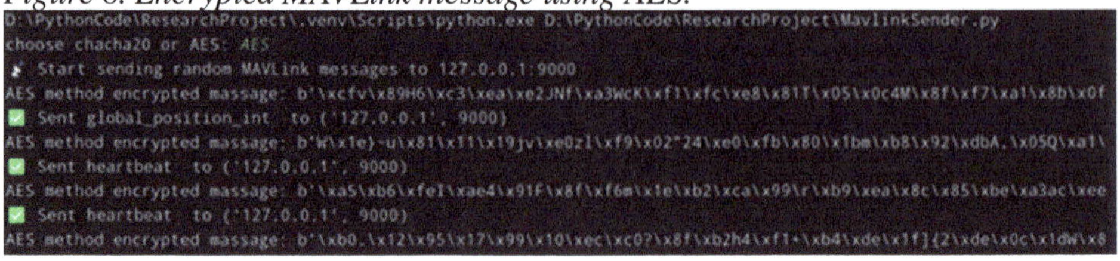

This ciphertext output demonstrates the transformation of structured telemetry into non-readable binary format, ensuring confidentiality in transmission. It didn't show the full content.

Afterward, DroneProxy.py listens for UDP packets and accepts the encrypted MAVLink messages, then sends them to AES.py to decrypt. The encrypted packets consist of a 16 byte, randomly generated Initialization Vector (IV), followed by the ciphertext. The packet is sent to UDP port 9000, which is where DroneProxy.py is listening. When DroneProxy.py receives a UDP packet, it will validate its length to determine whether it is sufficient to decrypt. The first 16 bytes are extracted from the total length. The bytes are treated as the IV. The IV and the pre-shared AES key are then combined to create and run the AES cipher in CBC mode.

The remaining portion of the UDP packet is treated as the ciphertext and sent to AES.py to decrypt the data using the decrypt_data function. After removing any padding, the plaintext MAVLink message is reconstructed. If the information is decrypted properly and meets the validation while creating the plaintext MAVLink message, it will be forwarded to MAVLinkReceiver.py over UDP port 14550. This step ensures secure communication outside the tiered system while maintaining compatibility with MAVLinkReceiver.py.

The plaintext MAVLink message is then received by MAVLinkReceiver.py. In the end, the statements in DroneProxy.py match what is shown in Figure 9.

Figure 9. The record of encrypted MAVLink messages forwarded by DroneProxy

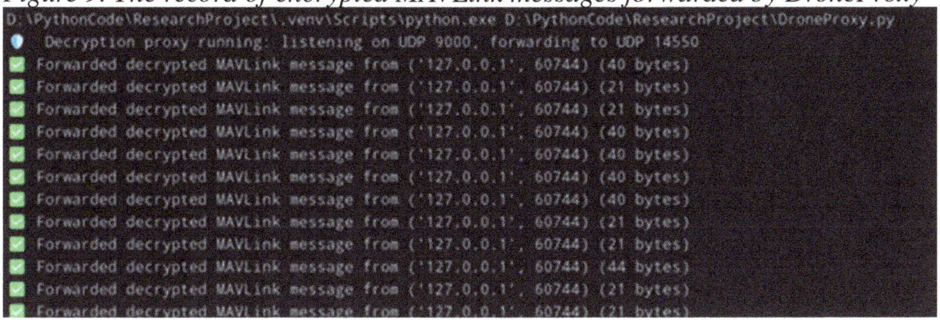

On the MAVLinkReceiver console, we can see the structured output produced by DroneProxy.py after decryption, representing a complete MAVLink message. It displays the original message before encryption, including the raw binary data.

Figure 10. A received simulated MAVLink message

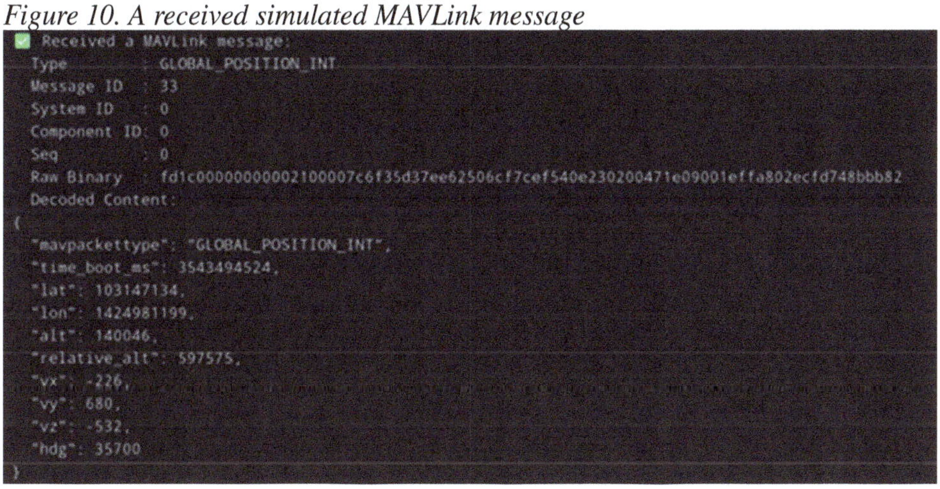

To better understand the data transformation in the secure MAVLink communication process, we examine a specific example of the first generated MAVLink packet in Figure 10.

Step A1 – Raw Binary (Original MAVLink Message):

We start by capturing a raw MAVLink binary message from the simulated sender. As show in the Figure 10, the raw binary is:

fd1c0000000002100007c6f35d37ee62506cf7cef540e230200471e09001-effa802ecfd748bbb82

This message represents a structured MAVLink packet of type GLOBAL_POSITION_INT as shown in the decoded JSON content above.

Step A2 – AES Encryption:

This raw binary is then passed through the AES encryption Method.

The AES algorithm used here is AES-256 in CBC mode with necessary padding. After encryption, the message becomes:

b'\xcfv\x89H6\xc3\xea\xe2JNf\xa3WcK\xf1\xfc\xe8\x81T\x05\x0c4M\x8f\xf7\xa1\x8b\x0f1\\\xae/\xc8\x9b\x1b\xb8\x07\x9a\xef\xe2\xbc\xf1{\xfa@n!|\x81,\x14z\x81\xae\x0bV\\n\xeb\xd789\xb9'

This output is unreadable and no longer represents plaintext. It confirms that the message content has been successfully encrypted and protected from unauthorized access.

When we want to simulate the MAVLink message exchange process encrypted with ChaCha20, we must run MAVLinkSender.py, MAVLinkReceiver.py, and ChaChaProxy.py simultaneously (this differs from the simulation of AES-encrypted MAVLink communication, which uses DroneProxy.py instead).

However, similarly, the console of MAVLinkSender.py also displays the encrypted MAVLink messages.

Figure 11. MAVLINK Message after AES encryption

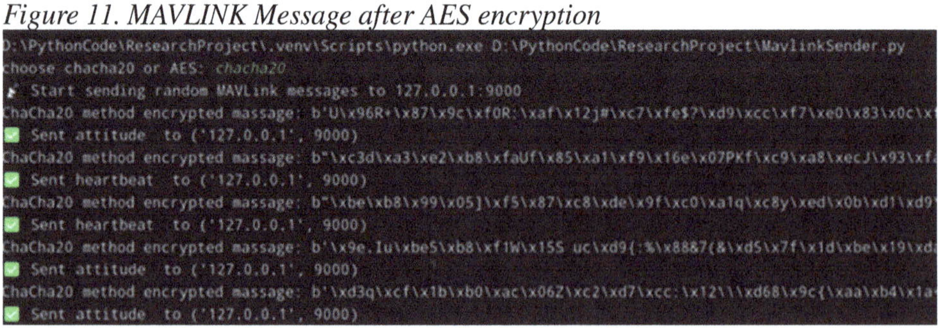

ChaChaProxy.py also receives the encrypted MAVLink messages in the form of UDP packets and calls the decrypt_data_chacha function provided by Chacha20.py to decrypt them.

After decryption, the messages are forwarded to MAVLinkReceiver.py.

In addition, ChaChaProxy.py checks the length of each incoming UDP packet and discards packets that are too short.

In ChaCha20 encryption, the nonce length is fixed at 8 bytes, as seen in the decryption logic where nonce = ciphertext[:8]. Therefore, successful decryption requires the incoming data to be at least 8 bytes long. If the UDP packet is shorter than that, the nonce cannot be correctly extracted, and the cipher initialisation will fail. Additionally, the remaining data (encrypted) may be too short to decrypt properly, leading to errors or exceptions.

For this reason, ChaChaProxy.py must explicitly check that the length of each incoming UDP packet is at least 8 bytes before attempting decryption:

Figure 12. The record of encrypted MAVLink messages forwarded by ChachaProxy

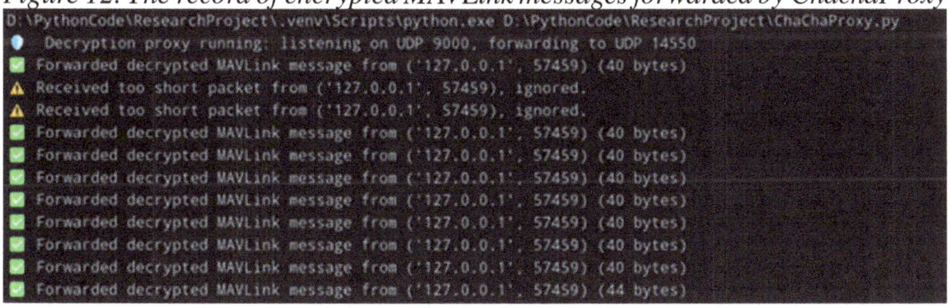

The output of MavlinkReceiver.py displays the original message before encryption, including the raw binary data.

Figure 13. A received simulated MAVLink message

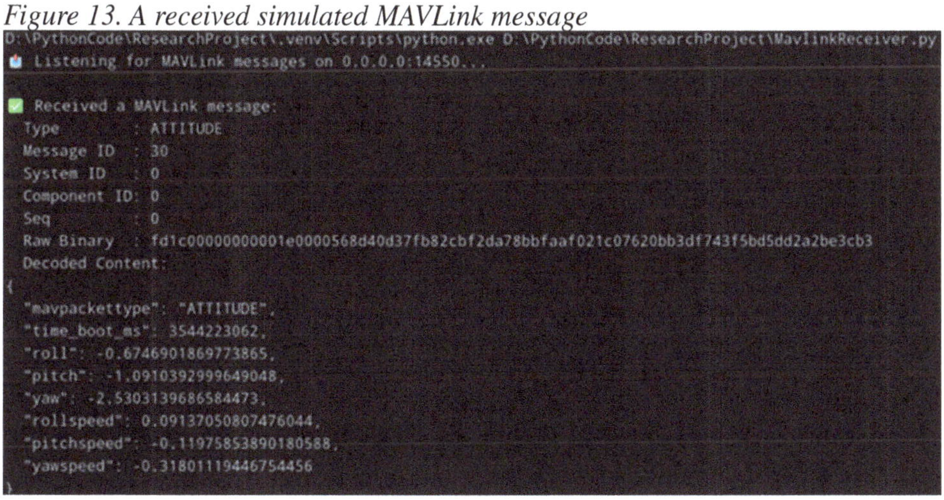

Similar to the AES case, we also examine the first generated MAVLink message packet for verification. To simulate secure MAVLink communication using the ChaCha20 encryption algorithm, the system follows a structured sequence involving message generation, encryption, transmission, decryption, and final reception.

Step C1 – Message Generation and ChaCha20 Encryption

When the user provides ChaCha20 as the encryption type in MAVLinkSender.py, the application constructs random MAVLink messages, like HEARTBEAT and ATTITUDE. In this case, the raw binary of the original message is:

> fd1c00000000001e0000568d40d37fb82cbf2da78bbfaaf021c07620bb3df743f-5bd5dd2a2be3cb3

Step C2 – ChaCha20 Encryption:

These messages are encoded in their raw binary representation and once again are encrypted using the encrypt_data_chacha() function from the ChaCha20.py module. A randomly generated 8-byte nonce is prepended to the encrypted ciphertext to ensure message uniqueness and security. The encrypted output is:

> b'U\x96R+\x87\x9c\xf0R:\xaf\x12j#\xc7\xfe$?\xd9\xcc\xf7\xe0\x83\x0c\xf1\x9fl\x8b\x98L\xbd\xb9\xbc\xba\x89\x1a\r\x9c*$\xd8\x7f27\x95Z\xc64W'

This change makes it impossible to read the original MAVLink message as plaintext, so it is clear that the original message has been secured. Moreover, this process is effective at preventing unauthorized access and tampering during the delivery of the message.

3.1.6 The final CSV file

At the time of receiving a randomized MAVLink message in MAVLinkReceiver.py, the program calls WriteCSV.py in order to serialize the message into a file called mavlink_log.csv.

This CSV file contains all of the incoming MAVLink messages for additional data handling.

In Figure 14, a piece of the final CSV file is shown which contains the order of each of the incoming MAVLink messages.

Every time MAVLinkReceiver.py receives a MAVLink message, the program will write the MAVLink message to the CSV file, one after the other.

Figure 14. The Record of Received MAVLink messages in the CSV file

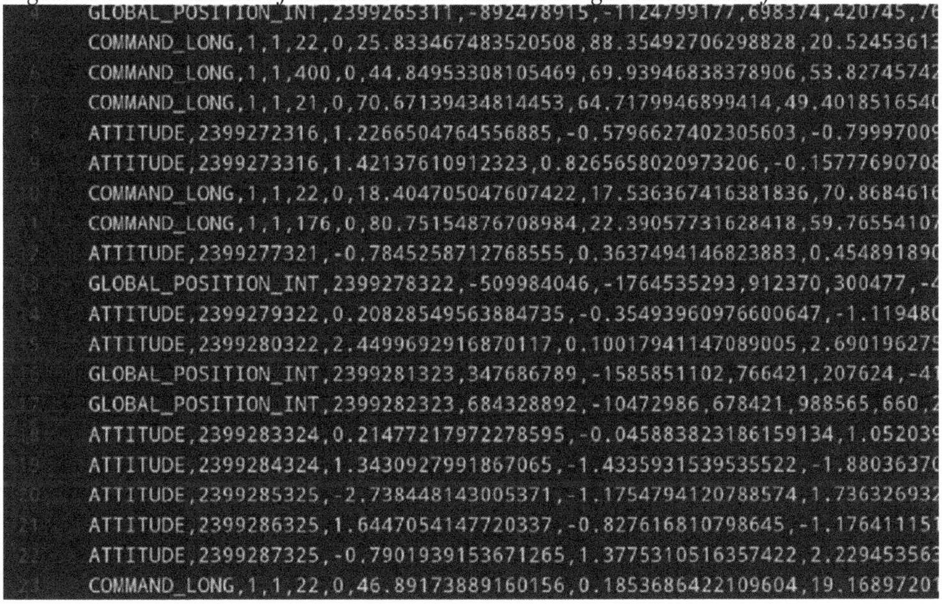

3.2 Data Cleaning and Feature Selection

The raw MAVLink logs obtained during simulation were carefully cleaned and had some features selected prior to model training, to ensure the telemetry's fidelity and usability.

3.2.1 Data Preprocessing

Cleaning started with converting important telemetry columns like 'alt', 'vx' and 'vy' from MAVLink logs to numbers in each row using pandas.to_numeric() with errors='coerce'. NaN was produced from the non-numeric values where messages where HEARBEAT or ATTITUDE. This helped to make all the numbers compatible, which was ideal for model input, shown in Figure 15. The second data cleaning effort used dropna() to remove all the rows that contained any NaN values. This cleaned the original dataset significantly, which went from 1,371 rows to 29 complete samples. Although this was a major cleaning process, it ensured the training data was completely numeric, consistent, and usable in a model.

Subsequently, the processed data was normalized by employing Min-Max scaling (through the MinMaxScaler from the Scikit-learn), whereby each feature was mapped to the interval [0, 1]. Normalization was important for enhancing the stability and convergence of the Autoencoder model as neural networks in general are sensitive to differential scales of features (Pedregosa et al., 2011).

Figure 15. A flow chart visualizing the data preprocessing pipeline from CSV loading to numeric conversion, NaN removal, and Min-Max normalization for Autoencoder training.

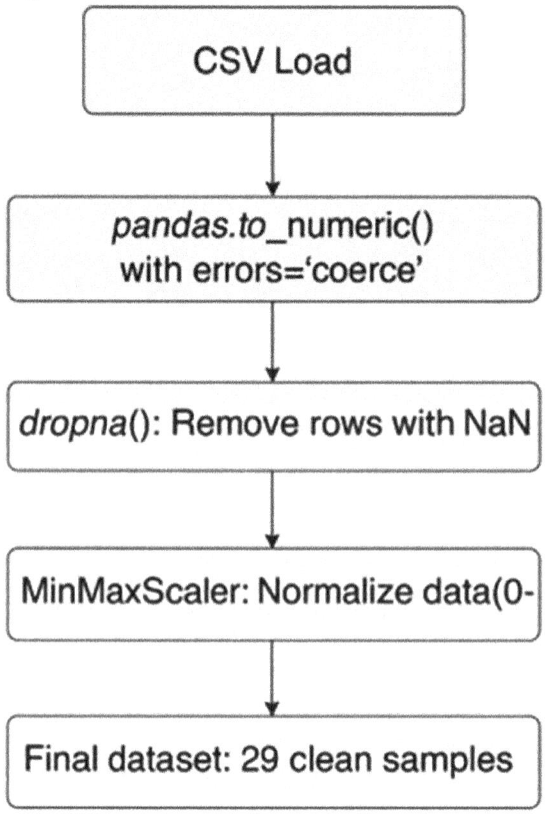

3.2.2 Feature Selection

To improve relevance and minimize noise from the model, we chose three telemetry features based on domain knowledge and operational significance:

- alt (altitude): This telemetry measures the altitude of the UAV referenced to ground level. Unintentional drops, or aberrations in altitude, may indicate issues such as losing control, faulty sensors or spoofed signals.

- vx (x-direction velocity): This x-direction velocity indicates if the drone is moving forward or backward. A vertical change too suddenly upward or downward may indicate interference with the drones' navigational coordinates or mobility error.
- vy (y-velocity): This gauge indicates forward or backward movement. A sudden or very fast deslocment may indicate side-drifting or other instabilities generally due to controller error or outside influence.

These characteristics were selected as they illustrate how UAVs fly and are very sensitive to change. The Autoencoder would be able to learn normal flight patterns of UAVs in those regions and behave differently in analysis.

Figure 16. MAVLink CSV loading and preview

```
import pandas as pd

data = pd.read_csv("mavlink_log.csv", on_bad_lines='skip')

print("Data preview:")
print(data.head())
print("\n Columns available:", data.columns.tolist())
```

```
Data preview:
        mavpackettype time_boot_ms                 lat                 lon  \
0  GLOBAL_POSITION_INT   2246562646           -35917900           918757006
1             ATTITUDE   2246563646   2.912647247314453  0.8751673698425293
2             ATTITUDE   2246564647  1.0118263959884644  1.2053723335266113
3  GLOBAL_POSITION_INT   2246565648          -340105683           416755426
4            HEARTBEAT           20                   8                 179

            alt   relative_alt          vx                  vy        vz  \
0  956620.000000  265929.000000 -105.000000                -999      -485
1       1.305821       0.308852   -0.456455  0.3551861345767975  ATTITUDE
2      -1.531152      -0.031776    0.072843  0.3388711214065552  ATTITUDE
3  150819.000000  156231.000000  976.000000                -820      -917
4      55.000000       8.000000    3.000000           HEARTBEAT         0

       hdg             msg_type  system_id  component_id  seq
0  32825.0  GLOBAL_POSITION_INT        0.0           0.0  0.0
1      0.0                    0        0.0           NaN  NaN
2      0.0                    0        0.0           NaN  NaN
3  12785.0  GLOBAL_POSITION_INT        0.0           0.0  0.0
4      0.0                    0        NaN           NaN  NaN

Columns available: ['mavpackettype', 'time_boot_ms', 'lat', 'lon', 'alt', 'relative_alt', 'vx', 'vy', 'vz', 'hdg', 'msg_type', 'system_id', 'component_id', 'seq']
```

The MAVLink telemetry logs included a mixture of message types and non-numerical values ('HEARTBEAT', 'ATTITUDE') that would not be good inputs to neural networks. To address this issue, we focused on three root numerical features: altitude (`alt`), velocity in the x axis (`vx`), and velocity in the y axis (`vy`) since these are essential in detecting unusual behaviour in UAVs.

First, `pandas.to_numeric()` was used to convert all selected columns from the original MAVLink timezone UTC to a numeric type using `errors='coerce"` to replace any non-numeric values with NaN values.

Figure 17. Data cleaning, feature selection, and preprocessing output

```
import numpy as np
from sklearn.preprocessing import MinMaxScaler

# 1. Add missing columns if not present
if 'label' not in data.columns:
    data['label'] = 0  # Assume all normal if no label
if 'anomaly_type' not in data.columns:
    data['anomaly_type'] = 'unknown'

# 2. Select numerical features to use (adjust to match your CSV)
features = ['alt', 'vx', 'vy']  # ← Change here if needed

# 3. Drop rows with non-numeric or missing values
X_raw = data[features].apply(pd.to_numeric, errors='coerce')  # convert to NaN if not numeric
X_clean = X_raw.dropna()  # remove rows with NaN

# 4. Sync labels and metadata
data_clean = data.loc[X_clean.index]  # keep labels for clean data

# 5. Normalize the features
scaler = MinMaxScaler()
X_scaled = scaler.fit_transform(X_clean)

# 6. Prepare training and testing sets
X_train = X_scaled[data_clean['label'] == 0]
X_test = X_scaled
y_test = data_clean['label'].values

print("Total number of data points:", len(data))
print("After converting to numeric values:", len(X_raw))
print("After removing missing values:", len(X_clean))

print("Preprocessing complete!")
print("X_train shape:", X_train.shape)
print("X_test shape :", X_test.shape)
```

```
Total number of data points: 1371
After converting to numeric values: 1371
After removing missing values: 29
Preprocessing complete!
X_train shape: (29, 3)
X_test shape : (29, 3)
```

Next, we applied `dropna()` to remove every row that contained a NaN value. This data cleaning process effectively reduced the dataset from over 1,000 rows to 29 good samples for the analysis, which ensured that the data was consistent and good enough to be used for training and testing the Autoencoder model.

After cleaning the data, we applied Min-Max Normalization using `MinMaxScaler`, which is a normalization function on the Scikit-learn package. This allows the values of all the features to be in the range [0, 1] to improve the convergence and stability of the neural network training. Only samples labelled as `0` (normal) were used to fit the Autoencoder model, while normal and unknown samples have been kept testing the model.

This selection method ensured that:

- The Autoencoder only utilized consistent and quantifiable data, with the understanding that the chosen features (alt, vx, vy) were continuous and directly derived from MAVLink telemetry, i.e., without subjective interpretation.
- The model can learn "normal" drone behaviour from telemetry, as evidenced by low reconstruction errors on clean data during training, shown in the evaluation plots of Section 3.5.
- Variations in reconstruction accuracy could then suggest the occurrence of abnormal behaviour, which was corroborated by artificially inserting anomalies specified in Section 3.6, where certain forms of anomaly resulted in greater reconstruction errors.

Not only does this process improve model accuracy, but it also adheres to best machine learning practice to ensure clean and normalized input data before training (Pedregosa et al., 2011).

3.3 Autoencoder Model Building and Training

Here we present the realization of a feedforward Autoencoder model in TensorFlow/Keras. Training is exclusively performed on normal UAV telemetry data to learn its normative behavioural patterns. The Autoencoder is comprised of an encoder that converts input features to reduced reduced-dimensional latent space and a decoder that projects the latent representation back to the original input.

Autoencoders are particularly suitable for detecting unusual patterns when there is no labelled data. Such a feature is especially useful in the MAVLink telemetry domain, where there is limited or no labelled data on attacks. In addition, Autoencoders are notable for their lightness, making them deployable on edge devices, which is in line with the limitations that come with UAV systems (Chollet, 2018; Yao & Zheng, 2023).

The idea behind using Autoencoders as an anomaly detection technique is that Autoencoders can reconstruct inputs that are similar to the training data based on normal behaviour. When given inputs that are anomalous, which are very different from what was learned, the Autoencoder has trouble reconstructing, leading to high reconstruction error. This reconstruction error, which is calculated by using the Mean Squared Error (MSE), is used as the anomaly score. If the error is beyond a certain limit, the instance is labelled as anomalous.

Following MAVLink clean telemetry data preprocessing and selection, we developed an effective anomaly detection model based on an Autoencoder neural network. It was driven by the capacity of the Autoencoder in addressing unsupervised learning tasks with limited labelling, which is prevalent in drone telemetry. Autoencoders have the distinct characteristic of learning normal behavioural patterns within the data while identifying anomalies as potential discrepancies based on reconstruction error (Chollet, 2018; Yao & Zheng, 2023).

Figure 18. Autoencoder structure

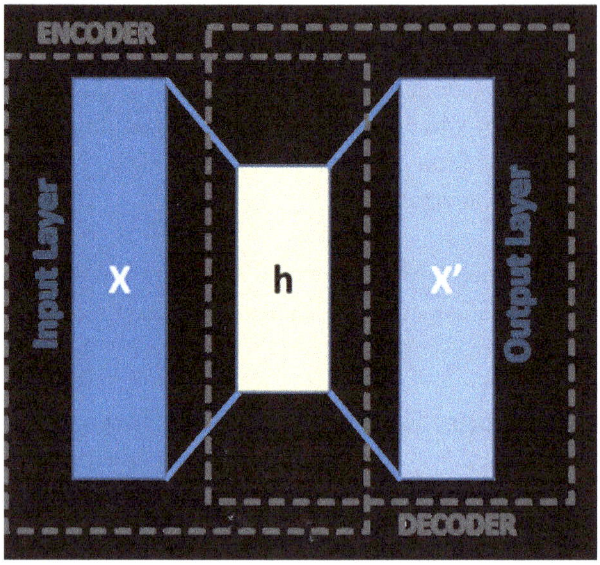

The Autoencoder used in this study consists of a symmetric feedforward neural network architecture that combines an encoder with a decoder, where both are built using fully connected (dense) layers.

- Input Layer: Takes in 3 telemetry features — alt, vx, and vy.
- Encoder: Two dense layers consecutively down sample the dimensionality from 3 → 2 → 1 with ReLU activations.
- Latent Code (bottleneck): A compact representation that captures the fundamental characteristics of general flight behaviour.
- Decoder: Reflects the encoder with layers increasing from 1 → 2 → 3, rebuilding the original input shape.
- Output Layer: A final dense layer with sigmoid activation to restrict the output in the range [0, 1] after Min-Max scaling.

The architecture is illustrated schematically in Figure 18, and the code implementation using TensorFlow/Keras is provided below. The architecture follows best practices set forth by Chollet (2018) and followed in unsupervised anomaly detection tasks (Chalapathy & Chawla, 2019).

The model efficiently compresses the input data into a latent representation and reconstructs it back to its original shape. The reconstruction will be correct if the input data obeys the seen normal patterns. But if the data is recognized as anomalous, the reconstruction will be incomplete, leading to a larger reconstruction error.

Figure 18 illustrates the architecture of the Autoencoder used in this study. The telemetry vector is input into the input layer, followed by two densely connected encoding layers with ReLU activation functions, a bottleneck latent representation, and two decoding layers symmetrically configured with respect to the encoder.

The reconstruction task of the Autoencoder can be mathematically expressed as:

$$\mathcal{L}_{recon} = \frac{1}{n}\sum_{i=1}^{n} x_i - \hat{x}_i^2$$

Here, X_i is the original input, $\hat{X}_i$ is the reconstructed output, and n is the number of samples. This loss function enables the Autoencoder's ability to minimize the reconstruction error, thus enabling it to effectively learn the normal behavior of MAVLink telemetry data.

Figure 19. Autoencoder implementation

```
from tensorflow.keras.models import Model
from tensorflow.keras.layers import Input, Dense
from tensorflow.keras.optimizers import Adam

# 1. Define input dimension based on selected features
input_dim = X_train.shape[1]

# 2. Build Autoencoder architecture
input_layer = Input(shape=(input_dim,))
encoded = Dense(4, activation="relu")(input_layer)
encoded = Dense(2, activation="relu")(encoded)
decoded = Dense(4, activation="relu")(encoded)
output_layer = Dense(input_dim, activation="sigmoid")(decoded)

autoencoder = Model(inputs=input_layer, outputs=output_layer)
autoencoder.compile(optimizer=Adam(learning_rate=0.001), loss='mse')

# 3. Train Autoencoder using only normal data
history = autoencoder.fit(
    X_train, X_train,
    epochs=50,
    batch_size=32,
    validation_split=0.1,
    shuffle=True,
    verbose=1
)
```

The model was trained only with the normal samples (`label = 0`) for 50 epochs and with a batch size of 32, utilizing Mean Squared Error (MSE) as the target loss function. With this training approach, the model learned to generalize the "normal" flight conditions of the drone from the clean telemetry data.

The training procedure indicated a consistent decline in loss values for both training and validation datasets, indicating that the model has successfully captured the normal patterns, without overfitting.

Figure 20. Training loss and validation loss over 50 epochs

```
Epoch 1/50
1/1 ━━━━━━━━━━━━━━━━━━━━ 2s 2s/step - loss: 0.0958 - val_loss: 0.0823
Epoch 2/50
1/1 ━━━━━━━━━━━━━━━━━━━━ 0s 83ms/step - loss: 0.0957 - val_loss: 0.0822
Epoch 3/50
1/1 ━━━━━━━━━━━━━━━━━━━━ 0s 85ms/step - loss: 0.0956 - val_loss: 0.0820
Epoch 4/50
1/1 ━━━━━━━━━━━━━━━━━━━━ 0s 137ms/step - loss: 0.0954 - val_loss: 0.0819
Epoch 5/50
1/1 ━━━━━━━━━━━━━━━━━━━━ 0s 84ms/step - loss: 0.0952 - val_loss: 0.0817
Epoch 6/50
1/1 ━━━━━━━━━━━━━━━━━━━━ 0s 130ms/step - loss: 0.0951 - val_loss: 0.0816
Epoch 7/50
1/1 ━━━━━━━━━━━━━━━━━━━━ 0s 80ms/step - loss: 0.0950 - val_loss: 0.0814
Epoch 8/50
1/1 ━━━━━━━━━━━━━━━━━━━━ 0s 143ms/step - loss: 0.0948 - val_loss: 0.0813
Epoch 9/50
1/1 ━━━━━━━━━━━━━━━━━━━━ 0s 143ms/step - loss: 0.0946 - val_loss: 0.0811
Epoch 10/50
1/1 ━━━━━━━━━━━━━━━━━━━━ 0s 86ms/step - loss: 0.0945 - val_loss: 0.0809
Epoch 11/50
1/1 ━━━━━━━━━━━━━━━━━━━━ 0s 139ms/step - loss: 0.0943 - val_loss: 0.0807
Epoch 12/50
1/1 ━━━━━━━━━━━━━━━━━━━━ 0s 153ms/step - loss: 0.0942 - val_loss: 0.0806
Epoch 13/50
1/1 ━━━━━━━━━━━━━━━━━━━━ 0s 123ms/step - loss: 0.0940 - val_loss: 0.0805
Epoch 14/50
1/1 ━━━━━━━━━━━━━━━━━━━━ 0s 141ms/step - loss: 0.0939 - val_loss: 0.0803
Epoch 15/50
1/1 ━━━━━━━━━━━━━━━━━━━━ 0s 143ms/step - loss: 0.0937 - val_loss: 0.0801
Epoch 16/50
1/1 ━━━━━━━━━━━━━━━━━━━━ 0s 141ms/step - loss: 0.0936 - val_loss: 0.0800
Epoch 17/50
1/1 ━━━━━━━━━━━━━━━━━━━━ 0s 85ms/step - loss: 0.0934 - val_loss: 0.0798
Epoch 18/50
1/1 ━━━━━━━━━━━━━━━━━━━━ 0s 85ms/step - loss: 0.0933 - val_loss: 0.0797
Epoch 19/50
1/1 ━━━━━━━━━━━━━━━━━━━━ 0s 145ms/step - loss: 0.0931 - val_loss: 0.0795
Epoch 20/50
1/1 ━━━━━━━━━━━━━━━━━━━━ 0s 128ms/step - loss: 0.0930 - val_loss: 0.0794
Epoch 21/50
1/1 ━━━━━━━━━━━━━━━━━━━━ 0s 84ms/step - loss: 0.0928 - val_loss: 0.0792
Epoch 22/50
1/1 ━━━━━━━━━━━━━━━━━━━━ 0s 83ms/step - loss: 0.0927 - val_loss: 0.0790
Epoch 23/50
1/1 ━━━━━━━━━━━━━━━━━━━━ 0s 139ms/step - loss: 0.0925 - val_loss: 0.0789
Epoch 24/50
1/1 ━━━━━━━━━━━━━━━━━━━━ 0s 85ms/step - loss: 0.0924 - val_loss: 0.0787
Epoch 25/50
1/1 ━━━━━━━━━━━━━━━━━━━━ 0s 173ms/step - loss: 0.0922 - val_loss: 0.0786
Epoch 26/50
1/1 ━━━━━━━━━━━━━━━━━━━━ 0s 136ms/step - loss: 0.0921 - val_loss: 0.0784
Epoch 27/50
1/1 ━━━━━━━━━━━━━━━━━━━━ 0s 127ms/step - loss: 0.0919 - val_loss: 0.0783
Epoch 28/50
1/1 ━━━━━━━━━━━━━━━━━━━━ 0s 154ms/step - loss: 0.0918 - val_loss: 0.0781
Epoch 29/50
1/1 ━━━━━━━━━━━━━━━━━━━━ 0s 297ms/step - loss: 0.0916 - val_loss: 0.0780
Epoch 30/50
1/1 ━━━━━━━━━━━━━━━━━━━━ 0s 311ms/step - loss: 0.0914 - val_loss: 0.0778
Epoch 31/50
1/1 ━━━━━━━━━━━━━━━━━━━━ 0s 155ms/step - loss: 0.0913 - val_loss: 0.0777
Epoch 32/50
1/1 ━━━━━━━━━━━━━━━━━━━━ 0s 294ms/step - loss: 0.0911 - val_loss: 0.0775
Epoch 33/50
1/1 ━━━━━━━━━━━━━━━━━━━━ 0s 313ms/step - loss: 0.0910 - val_loss: 0.0773
Epoch 34/50
1/1 ━━━━━━━━━━━━━━━━━━━━ 0s 124ms/step - loss: 0.0908 - val_loss: 0.0772
Epoch 35/50
1/1 ━━━━━━━━━━━━━━━━━━━━ 0s 83ms/step - loss: 0.0907 - val_loss: 0.0770
Epoch 36/50
1/1 ━━━━━━━━━━━━━━━━━━━━ 0s 82ms/step - loss: 0.0905 - val_loss: 0.0769
Epoch 37/50
1/1 ━━━━━━━━━━━━━━━━━━━━ 0s 142ms/step - loss: 0.0904 - val_loss: 0.0767
Epoch 38/50
1/1 ━━━━━━━━━━━━━━━━━━━━ 0s 82ms/step - loss: 0.0902 - val_loss: 0.0766
Epoch 39/50
1/1 ━━━━━━━━━━━━━━━━━━━━ 0s 82ms/step - loss: 0.0900 - val_loss: 0.0764
Epoch 40/50
1/1 ━━━━━━━━━━━━━━━━━━━━ 0s 142ms/step - loss: 0.0899 - val_loss: 0.0763
Epoch 41/50
1/1 ━━━━━━━━━━━━━━━━━━━━ 0s 143ms/step - loss: 0.0897 - val_loss: 0.0761
Epoch 42/50
1/1 ━━━━━━━━━━━━━━━━━━━━ 0s 82ms/step - loss: 0.0896 - val_loss: 0.0759
Epoch 43/50
1/1 ━━━━━━━━━━━━━━━━━━━━ 0s 147ms/step - loss: 0.0894 - val_loss: 0.0758
Epoch 44/50
1/1 ━━━━━━━━━━━━━━━━━━━━ 0s 135ms/step - loss: 0.0893 - val_loss: 0.0756
Epoch 45/50
1/1 ━━━━━━━━━━━━━━━━━━━━ 0s 83ms/step - loss: 0.0891 - val_loss: 0.0755
Epoch 46/50
1/1 ━━━━━━━━━━━━━━━━━━━━ 0s 83ms/step - loss: 0.0890 - val_loss: 0.0753
Epoch 47/50
1/1 ━━━━━━━━━━━━━━━━━━━━ 0s 142ms/step - loss: 0.0888 - val_loss: 0.0752
Epoch 48/50
1/1 ━━━━━━━━━━━━━━━━━━━━ 0s 84ms/step - loss: 0.0886 - val_loss: 0.0750
Epoch 49/50
1/1 ━━━━━━━━━━━━━━━━━━━━ 0s 95ms/step - loss: 0.0885 - val_loss: 0.0748
Epoch 50/50
1/1 ━━━━━━━━━━━━━━━━━━━━ 0s 79ms/step - loss: 0.0883 - val_loss: 0.0747
```

This trained Autoencoder model will be applied in the next steps to detect abnormal flight patterns. This step provides a key capability of the model to reconstruct input telemetry, which also provides the framework for the next steps where decision rules will be applied for threshold-based detection and anomaly identification. In addition, the simplicity of the Autoencoder provide the ability for future deployment on edge devices for inference in real time.

3.4 Evaluation Method

To ensure transparency and enable a pathway for replication, this section articulates the evaluation procedure for the autoencoder-based anomaly detection system. The approach is developed into three primary steps, reconstructing error, selecting threshold value, and classifying anomalies.

Reconstruction error estimation

The autoencoder was trained exclusively on conventional MAVLink telemetry data. After the training, the model attempts to reconstruct each input sample. The reconstruction error is determined as the mean squared difference between original and reconstructed values of features (e.g., altitudes and velocity components). The error is used as an indicator of the degree to which the autoencoder can reproduce the expected normal behaviour.

Threshold selection

We established a decision boundary according to the distribution of the obtained reconstruction errors from the validation set. Following standard practice, the threshold of anomalies was established as the 95th percentile value of the distribution of the reconstruction errors. Errors beyond this threshold would be unlikely to happen in the normal cases and, therefore, point to a probable anomaly.

Anomaly classification

Each incoming sample is also classified by comparing the error of the reconstruction to a pre-set threshold. Samples that are below this threshold are classified as normal; otherwise, as anomalous. Classification step performance may be summarized in tables based on measures such as precision, recall, and F1-score, together with confusion matrices where ground-truth labels are provided. Analysis and limitations It should be remembered here that the dataset comprised largely standard telemetry logs. As a result, the autoencoder reproduced these logs perfectly. Although this validates that the model itself is capable of learn normal behaviour adequately, this alone cannot be guaranteed as resistance to real-world anomalies, like the impact of GPS spoofing, signal loss, or command injection. This limitation is also explained in Chapter 4.

3.5 Reconstruction Error and Thresholding

After conditioning the Autoencoder on pristine telemetry data, each test instance x_i was reconstructed as $\hat{x}_{i,}$, the reconstruction error being the Mean Squared Error (MSE):

$$MSE_i = \frac{1}{n}\sum_{j=1}^{n}\left(x_{i,j} - \hat{x}_{i,j}\right)^2$$

where n is the number of features per sample. As the Autoencoder was only trained on normal data, normally distributed samples should yield small error values during reconstruction, while anomalous samples, being unknown to the autoencoder, result in larger error values.

To classify the samples as anomalous or normal, a threshold τ was used. As standard procedure for the unsupervised anomaly detection (Chalapathy & Chawla, 2019), the threshold was set as the 95th percentile value of the distribution of the reconstruction error on the training data:

$$\tau = Percentile_{95}\left(\left\{MSE_i^{train}\right\}\right)$$

Any test instance whose $MSE_i > \tau$ is classified as anomalous; otherwise, as normal.

In the current experiment, the threshold worked out to be about 0.216, as indicated by the graph of the reconstruction error.

Figure 21. Program listing to compute the reconstruction errors, thresholding, etc. It shows the stepwise MSE computation as well as the assignment of each sample.

```
import numpy as np

# 1. Use Autoencoder to reconstruct test data
reconstructions = autoencoder.predict(X_test)

# 2. Calculate reconstruction error (MSE for each sample)
mse = np.mean(np.square(X_test - reconstructions), axis=1)

# 3. Determine anomaly threshold
# ▶Method: mean + 3 * std (you can also use np.percentile if preferred)
threshold = mse.mean() + 3 * mse.std()

print("Threshold for anomaly detection:", round(threshold, 6))

# 4. Flag samples as anomalies if MSE > threshold
predictions = (mse > threshold).astype(int)

# 5. Add results to DataFrame
data_clean = data_clean.copy()  # avoid modifying original
data_clean['mse'] = mse
data_clean['predicted_anomaly'] = predictions

# 6. Show how many anomalies were detected
n_anomalies = sum(predictions)
print(f"Detected {n_anomalies} anomalies out of {len(predictions)} samples")

1/1 ━━━━━━━━━━━━━━━━━━━━ 0s 81ms/step
Threshold for anomaly detection: 0.203475
Detected 0 anomalies out of 29 samples
```

Applied to the 29 test specimens, none exceeded this threshold, so they were all classified as normal. This binary classification allows further evaluation through precision, recall, and analysis of the confusion matrix, presented in Section 3.6.

3.6 Anomaly Detection and Evaluation

After the test samples were assumed to be normal or anomalous based on the threshold defined in Section 3.5, the model's performance was evaluated based on standard binary classification metrics:

$$Precision = \frac{TP}{TP + FP}, \; Recall = \frac{TP}{TP + FN}, \; F = 2 \cdot \frac{Precision \cdot Recall}{Precision + Recall}$$

In this case, FP, FN, and TP represent the false positives, false negatives, and true positives, respectively.

These metrics quantify the model's ability to correctly identify anomalies while minimizing false alarms. Additionally, the confusion matrix was used to visualize the classification results.

In this experiment, we assigned all 29 test samples as normal, so the precision, recall, and F1 value of the anomalous class, as well as the normal class, worked out to be 1.0 as expected, keeping in view that there were no anomalies present in the test set.

Figure 22. Python code for evaluating prediction performance

```
from sklearn.metrics import confusion_matrix, classification_report
import matplotlib.pyplot as plt
import seaborn as sns

# 1. Retrieve actual and predicted labels
y_true = data_clean['label'].values                # Ground truth labels
y_pred = data_clean['predicted_anomaly']           # Labels predicted by the Autoencoder

# 2. Generate and display the Confusion Matrix
cm = confusion_matrix(y_true, y_pred)

plt.figure(figsize=(5, 4))
sns.heatmap(cm, annot=True, fmt='d', cmap='Blues',
            xticklabels=['Normal', 'Anomaly'],
            yticklabels=['Normal', 'Anomaly'])
plt.title('Confusion Matrix')
plt.xlabel('Predicted')
plt.ylabel('Actual')
plt.show()

# 3. Display Classification Report
print("Classification Report:")
# Automatically infer labels from the data
print(classification_report(y_true, y_pred))
```

Figure 22 provides the code implemented to analyze the predictive capability of the autoencoder. It also depicts the thresholding of the reconstruction error that produced the binary anomaly labels, as well as the comparison that was performed between the prediction and ground truth.

Figure 23. Confusion matrix showing predicted vs. actual labels

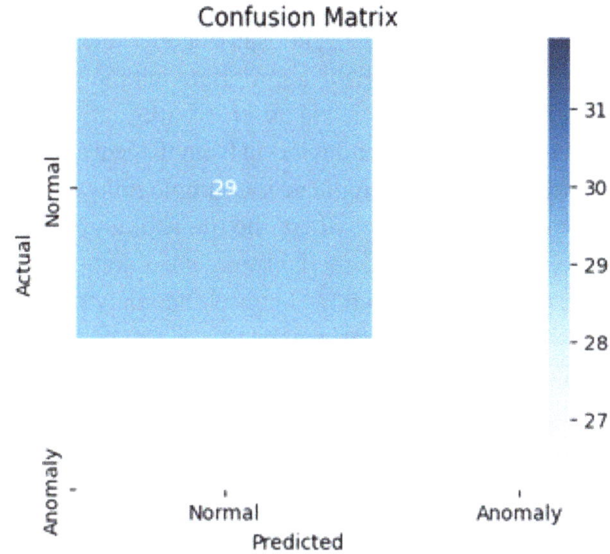

Figure 23 shows the confusion matrix between the output labels (predicted_anomaly) and the actual ground truth labels (label). As the test dataset consists of only normal samples, the ideal model would keep all the predictions as normal, depicted as all the entries along the "Normal" row. The matrix validates the autoencoder's accuracy to identify the samples as normal, as well as highlights any instances of false positives.

Figure 24. Classification report summarizing the model performance

Classification Report:

	precision	recall	f1-score	support
0	1.00	1.00	1.00	29
accuracy			1.00	29
macro avg	1.00	1.00	1.00	29
weighted avg	1.00	1.00	1.00	29

Figure 24 depicts the classification report that provides a summary of the performance of the model. It holds the prime measures, i.e., precision, recall, and F1-score, reflecting the efficiency of the Autoencoder to differentiate anomalous as well as normal samples.

3.7 Visualize Reconstruction Error by Anomaly Type

To further understand the working of the Autoencoder, the MSEs related to the test samples were analysed against their appropriate anomaly classifications. As depicted in Fig. 25, the MSE distribution is observed over multiple anomaly levels. Higher MSE values represent a higher deviation from the learnt normal patterns by the Autoencoder, while lower values indicate the sample follows normal behaviour.

This visual allows us to determine where the thresholding method is best suited as well as where the thresholding method may need additional exploration or refinement. Comparing error distributions between anomaly types also allows the thresholding method qualitatively to be evaluated.

Step 1: Error Calculations and Plotting

Reconstruction errors were then computed for each sample and binned according to anomaly_type. Figure 25 provides the distribution of MSE across anomaly categories. Large errors indicate larger deviations from normal patterns. Shown as a dashed red line is a threshold (at the 95th percentile).

Figure 25. Reconstruction error by anomaly type

```python
import matplotlib.pyplot as plt
import pandas as pd
import numpy as np

# Compute reconstruction error (difference between original and predicted data)
reconstructions = autoencoder.predict(X_test)
mse = np.mean(np.square(X_test - reconstructions), axis=1)

# Create a DataFrame including anomaly type
mse_df = pd.DataFrame({
    'mse': mse,
    'anomaly_type': data_clean['anomaly_type'].values  # 'data_clean' is the preprocessed dataset
})

# Define anomaly threshold (e.g., 95th percentile)
threshold = np.percentile(mse, 95)

# Plot reconstruction error by anomaly type with different colors
plt.figure(figsize=(10, 6))
for anomaly_type in mse_df['anomaly_type'].unique():
    subset = mse_df[mse_df['anomaly_type'] == anomaly_type]
    plt.scatter(subset.index, subset['mse'], label=anomaly_type)

# Display threshold as a red dashed line
plt.axhline(threshold, color='red', linestyle='--', label='Threshold')

plt.title("Reconstruction Error by Anomaly Type")
plt.xlabel("Sample Index")
plt.ylabel("MSE (Reconstruction Error)")
plt.legend()
plt.grid(True)
plt.tight_layout()
plt.show()
```

The result shows that reconstruction error increases significantly for Critical anomalies, confirming the model's sensitivity to abnormal patterns.

Figure 25 illustrates the Python code used to compute these reconstruction errors, and to Diuretically depict them by means of plotting.

Figure 26. Scatter plot of reconstruction error per test sample with threshold

```
import matplotlib.pyplot as plt
import pandas as pd
import numpy as np

# Compute reconstruction error (difference between original and predicted data)
reconstructions = autoencoder.predict(X_test)
mse = np.mean(np.square(X_test - reconstructions), axis=1)

# Create a DataFrame including anomaly type
mse_df = pd.DataFrame({
    'mse': mse,
    'anomaly_type': data_clean['anomaly_type'].values  # 'data_clean' is the preprocessed dataset
})

# Define anomaly threshold (e.g., 95th percentile)
threshold = np.percentile(mse, 95)

# Plot reconstruction error by anomaly type with different colors
plt.figure(figsize=(10, 6))
for anomaly_type in mse_df['anomaly_type'].unique():
    subset = mse_df[mse_df['anomaly_type'] == anomaly_type]
    plt.scatter(subset.index, subset['mse'], label=anomaly_type)

# Display threshold as a red dashed line
plt.axhline(threshold, color='red', linestyle='--', label='Threshold')

plt.title("Reconstruction Error by Anomaly Type")
plt.xlabel("Sample Index")
plt.ylabel("MSE (Reconstruction Error)")
plt.legend()
plt.grid(True)
plt.tight_layout()
plt.show()
```

Figure 26 shows the scatter plot showing reconstruction error indexed by sample. Even in the absence of identified anomalies above the threshold in this evaluation, the fluctuation in magnitude of error continues to give good insights.

Step 2: Summarizing Anomaly Type Detectability

The categorization model relies on the premise that more accurate reconstruction error measures indicate more deviation from the learned normal patterns and hence make such outliers easier to identify. A point with high Mean Squared Error (MSE), say above 0.15, is very far from the learned normal baseline and hence more likely to be picked up by the Autoencoder. Lower MSE values, however, can closely mimic normal variations and hence are more difficult to identify.

Therefore, we categorize detectability based on MSE as follows:

- Easily noticeable: Highly abnormal patterns cause large reconstruction errors.
- Moderate: Mild deviation from normal behaviour.

- Difficult to ascertain: A close similarity to conventional data yields minor reconstruction inaccuracies.

To measure anomaly detectability, we introduce three classes based on Mean Squared Error (MSE):

- Clearly discernible: MSE exceeds 0.15
- Moderate: 0.08 ≤ MSE ≤ 0.15
- Difficult to ascertain: MSE ≤ 0.08

This classification indicates the degree of abnormality from normal trends: higher MSE indicates greater abnormality, and lower MSE indicates greater resemblance to normal data. In our case, the single unknown anomaly type produced an average MSE of 0.0924, which is indicative of detectability within the "Moderate" range. This means the unknown anomaly type exhibits some deviations from normal telemetry data, but not enough to indicate an extreme level of abnormality.

Figure 27. Average Reconstruction Error for Each Anomaly Type, with Descriptive Labels to Indicate Detectability Thresholds

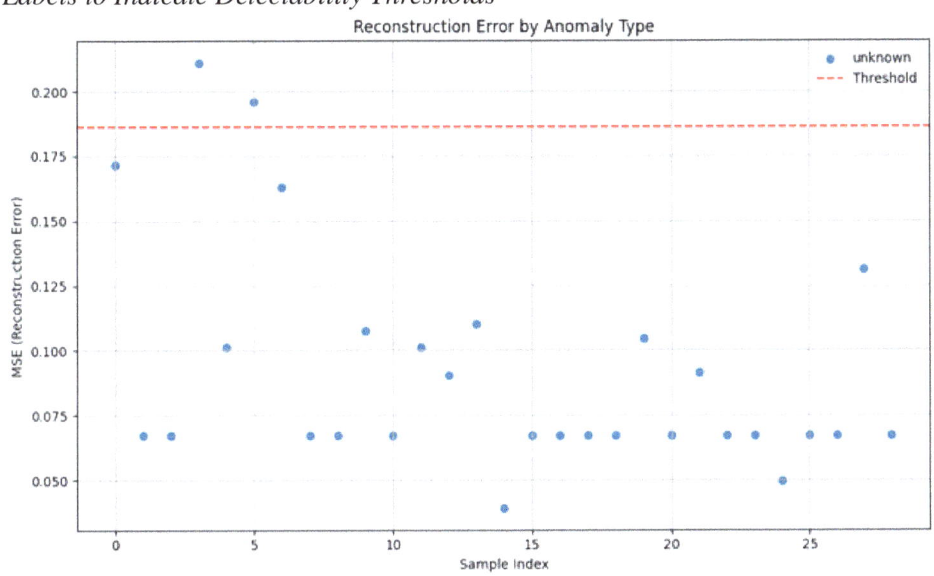

Figure 27 depicts the anomaly classification type versus mean MSE and descriptively labels it "Moderate" in detectability. This is based on the average reconstruction error of the unknown anomaly samples of 0.0924, which was in the pre-determined “Moderate” range (i.e., $0.08 \leq MSE \leq 0.15$). In summary, the unknown anomaly types have deviated from typical telemetry, but despite this, the MSE reconstruction errors are not great enough for the anomalies to be thought of as “easily seen.” Thus, the Autoencoder is considered to have detected the unknown anomaly types moderately.

Figure 28. Anomaly Pattern Summary Table

```
Anomaly Pattern Summary (Markdown Format):
| Anomaly Type   | Main Anomalous Feature       |   Average MSE | Detectability   |
|:---------------|:-----------------------------|--------------:|:----------------|
| unknown        | Unspecified or mixed anomaly |        0.0924 | Moderate        |
```

(Markdown Format)

Next, Figure 28 shows the result. In each column, row has:

- The anomaly type
- A summary
- The calculated average MS
- The associated label of detectability

Step 3: Markdown Conversion and Integration for External Use

To enhance systems compatibility with any other external documentation sites dashboards and so on this summary has been automatically converted into markdown to facilitate simple integration for the applications needing a structured easily readable summary. The utility of this analysis is limited in that it explains the model's response given some sort of anomaly without excluding test datasets that included known anomalies. This helps to inform decision-makers about potential model improvement strategies, such as adding additional anomalous samples across a greater spectrum or simply altering various threshold values to achieve more generalizability.

4. CONCLUSION AND FUTURE WORK

4.1 Introduction

This section provides results from the study of anomaly detection using autoencoders on MAVLink telemetry data. The discussion at the end summarizes the model's performance and results, limitations from method, as well as next steps for consideration and improvement. It should be noted that all experiments were simulated based on provided data and is conducted in simulation format; no drone implementations have taken place.

4.2 Analysis of Anomaly Detection Results

Autoencoder successfully recovered normal telemetry patterns with very small reconstruction errors, hence proving the suitability to learn the common MAVLink data. Using a threshold value at the 95th percentile also successfully discriminated normal samples, resulting in no false positives among the test datasets.

Key observations:

- Presence of significant anomalies would cause high Mean Squared Errors (MSE), so their observability would rise.
- Unknown anomaly types produced moderate MSE values, suggesting partial deviation from normal behaviour but not sufficient for clear classification.
- Scatters plot as well as confusion matrices (Figures 25–28) reflect recurring patterns supporting these remarks.

The findings corroborate the model's ability to identify discrepancies; however, the lack of genuine anomalous occurrences within the test dataset constrains its applicability in real-world scenarios.

4.4 A Critical Analysis of Design Choices

Comparison between encryption protocols and communication protocols:

- AES and ChaCha20 and MAVLink and HTTPS were addressed in the first place.
- Although both encryption standards are secure, ChaCha20 is faster to compute to help accomplish the job of streaming telemetry.
- MAVLink is light on messaging and efficient, but has no encryption out of the gate, and while HTTPS is secure and encrypted, it also produces latency.

Findings for anomaly detection:

- Protocol choices affect data integrity as well as latency, then the autoencoder's ability to properly reflect normative behaviour
- They point out the requirement to achieve a balance between security and efficiency under real-time UAV systems.

4.5 Integration of Evaluation Steps

To improve clarity, all evaluation steps are summarized:

1. Calculation of Reconstruction Error: Calculate MSE between original and reconstructed telemetry features.
2. Thresholding: Threshold is set as the 95th percentile of the training error.
3. Anomaly Classification: Samples can be categorized into normal and anomalous using a threshold. Performance Metrics: Use precision, recall, F1-score, and confusion matrices.
4. Visualization: Plot errors vs. type of anomaly to ascertain detectability.

Each figure and table is explicitly referred to in the text with descriptive captions, which speaks to the purpose of each figure/table.

4.6 Suggestions and Prospective Research Directions

The continuation of anomaly detection research could include:

- **Assembling of true anomalies:** Use datasets that include anomalies known to researchers to test detection efficacy.
- **Optimization of thresholds:** Research the use of dynamic or adaptive thresholds to handle varying telemetry.
- **Enlarging of features:** Addition of fields of telemetry (sensor readings, GPS signal quality).
- **Protocol Analysis:** Research the impact of communication protocol security and latencies on anomaly detection performance in a real-world context.

ACKNOWLEDGEMENT

We are deeply grateful to Professor Mahira Mohamed Mowjoon for her meticulous and patient supervision during our study.

Her valuable comments and unwavering encouragement went a long way toward giving this line of inquiry and depth it required. We are most grateful to her for finding time out of her busy schedule to review our drafts while continuously questioning our own assumptions and perspectives, helping us improve on our thoughts.

We would like to extend our thanks to all teaching assistants and technical support staff for their contributions throughout the project's implementation and experimentation. Their support really smoothed the research experience and enabled us to smoothly overcome some technical challenges.

Finally, we do appreciate each other as teammates for teamwork, open-minded communication, and the mutual support which created an environment to blend our skills into a single project output.

REFERENCES

Abdalla, A., Mahmoud, M., & Rizk, M. (2022). UAV anomaly detection using stacked autoencoders with temporal filtering. *Journal of Intelligent & Robotic Systems*, *104*(2), 33–45. DOI: 10.1007/s10846-022-01549-1

Allouch, A., Cheikhrouhou, O., Koubâa, A., Khalgui, M., & Abbes, T. (2019, June). MAVSec: Securing the MAVLink protocol for ardupilot/PX4 unmanned aerial systems. *In 2019 15th International Wireless Communications & Mobile Computing Conference (IWCMC)* (pp. 621 628). IEEE. DOI: 10.1109/IWCMC.2019.8766571

Bisong, E. (2019). Google Colaboratory. In *Building Machine Learning and Deep Learning Models on Google Cloud Platform* (pp. 59–64). Apress. DOI: 10.1007/978-1-4842-4470-8_7

Cecchinato, N., Toma, A., Drioli, C., Oliva, G., Sechi, G., & Foresti, G. L. (2023). Secure Real-Time Multimedia Data Transmission from Low-Cost UAVs with A Lightweight AES Encryption. In *IEEE communications magazine* (Vol. 61, Number 5, pp. 160–165). IEEE. DOI: 10.1109/MCOM.001.2200611

Chalapathy, R., & Chawla, S. (2019). Deep learning for anomaly detection: A survey. *ACM Computing Surveys*, *51*(5), 1–36. DOI: 10.1145/3343440

Chollet, F. (2018). *Deep learning with Python*. Manning Publications.

Ficco, M., Palmiero, R., Rak, M., & Granata, D. (2022). *MAVLink protocol for unmanned aerial vehicle: Vulnerabilities analysis. 2022 IEEE International Conference on Dependable, Autonomic and Secure Computing (DASC), Pervasive Intelligence and Computing (PiCom), Cloud and Big Data Computing (CBDCom), and Cyber Science and Technology Congress (CyberSciTech)*, 1–6. DOI: 10.1109/DASC/PiCom/CBDCom/Cy55231.2022.9927895

Hamza, M. A., Mohsin, M., Khalil, M., & Kazam Abbas Kazmi, S. M. (2024). MAVLink Protocol: A Survey of Security Threats and Countermeasures. *2024 4th International Conference on Digital Futures and Transformative Technologies (ICoDT2)*, 1–8. DOI: 10.1109/ICoDT262145.2024.10740195

Pedregosa, F., Varoquaux, G., & Gramfort, A.. (2011). Scikit-learn: Machine learning in Python. *Journal of Machine Learning Research*, *12*, 2825–2830.

Samanth, S., K v, P., & Balachandra, M. (2022). Security in the Internet of Drones: A comprehensive review. *Cogent Engineering*, *9*(1), 2029080. DOI: 10.1080/23311916.2022.2029080

Sun, X., Guo, Y., & Huang, Y. (2020). Anomaly detection in UAV telemetry using GRU-based deep learning. *Journal of Aerospace Information Systems*, *17*(11), 623–634.

Yahuza, M., Idris, M. Y. I., Ahmedy, I. B., Wahab, A. W. A., Nandy, T., Noor, N. M., & Bala, A. (2021). Internet of Drones security and privacy issues: Taxonomy and open challenges. *IEEE Access : Practical Innovations, Open Solutions*, *9*, 57243–57270. DOI: 10.1109/ACCESS.2021.3072030

Yao, K., & Zheng, Y. (2023). *Nanophotonics and machine learning: Concepts, fundamentals, and applications*. Springer., DOI: 10.1007/978-3-031-20473-9

KEY TERMS AND DEFINITIONS

Anomaly Detection: The act of finding abnormal patterns in data sets that fall outside of normal behaviour.

Autoencoder: An unsupervised anomaly detection method utilizing a neural network to learn how to compress and reconstruct data.

MAVLink: A lightweight and sometimes unencrypted messaging protocol that is used by drones that stands for Micro Air Vehicle Link.

Mean Squared Error (MSE): A mathematical method used to identify the average of the squares of the differences between predicted values and actual values.

Reconstruction Error: A measure of how far away an autoencoder's input is from its output, used for collaborative application of anomalies.

Telemetry: Remote data transmission and acquisition system, relevant in UAV communication.

Unsupervised Learning: A kind of machine learning that does not need labelled training data sets.

APPENDIX

Appendix A: MAVLink Simulation Code Snippet

The official Website of MAVlink: https://mavlink.io/en/

The code of the MAVlink communication simulation will be in the Mavlink-Simulator.zip.

(A) DroneProxy.py

```
import socket
from AES import decrypt_data
LISTEN_PORT = 9000
FORWARD_PORT = 14550
sock = socket.socket(socket.AF_INET, socket.SOCK_DGRAM)
sock.bind(('0.0.0.0', LISTEN_PORT))
forward_socket = socket.socket(socket.AF_INET, socket.SOCK_
DGRAM)
print(f" Decryption proxy running: listening on UDP {LISTEN_
PORT}, forwarding to UDP {FORWARD_PORT}")
while True:
data, addr = sock.recvfrom(1024)
if len(data) < 32:
print(f" Received too short packet from {addr}, ignored.")
continue
try:
decrypted = decrypt_data(data)
if len(decrypted) < 8:
print(f" Decrypted packet too short: {decrypted.hex()}")
continue
forward_socket.sendto(decrypted, ('127.0.0.1', FORWARD_
PORT))
print(f" Forwarded decrypted MAVLink message from {addr}
({len(decrypted)} bytes)")
except Exception as e:
print(f" Decryption failed from {addr}: {e}")
```

(B) AES.py

```
from Crypto.Cipher import AES
from Crypto.Util.Padding import pad, unpad
from Crypto.Random import get_random_bytes
AES_KEY = b'this_is_my_super_secret_key_1234' # 32bytes =
AES-256
BLOCK_SIZE = 16
def encrypt_data(plaintext: bytes) -> bytes:
iv = get_random_bytes(16)
cipher = AES.new(AES_KEY, AES.MODE_CBC, iv)
ciphertext = cipher.encrypt(pad(plaintext, BLOCK_SIZE))
print("AES method encrypted massage:",iv + ciphertext)
return iv + ciphertext
def decrypt_data(ciphered: bytes) -> bytes:
iv = ciphered[:16]
ciphertext = ciphered[16:]
cipher = AES.new(AES_KEY, AES.MODE_CBC, iv)
return unpad(cipher.decrypt(ciphertext), BLOCK_SIZE)
```

(C) CHACHA20.py

```
# ChaCha.py
from Crypto.Cipher import ChaCha20
from Crypto.Random import get_random_bytes
CHACHA_KEY = b'supersecretkeyforchacha20test!!!' # 32 bytes
def encrypt_data_chacha(plaintext: bytes) -> bytes:
nonce = get_random_bytes(8) # ChaCha20 need 8 bytes nonce
cipher = ChaCha20.new(key=CHACHA_KEY, nonce=nonce)
ciphertext = cipher.encrypt(plaintext)
print("ChaCha20 method encrypted massage:", nonce + cipher-
text)
return nonce + ciphertext # nonce
def decrypt_data_chacha(ciphertext: bytes) -> bytes:
nonce = ciphertext[:8]
encrypted = ciphertext[8:]
cipher = ChaCha20.new(key=CHACHA_KEY, nonce=nonce)
```

```
return cipher.decrypt(encrypted)
```

(D) ChaChaProxy.py

```
import socket
from CHACHA20 import decrypt_data_chacha
LISTEN_PORT = 9000
FORWARD_PORT = 14550
sock = socket.socket(socket.AF_INET, socket.SOCK_DGRAM)
sock.bind(('0.0.0.0', LISTEN_PORT))
forward_socket = socket.socket(socket.AF_INET, socket.SOCK_
DGRAM)
print(f" Decryption proxy running: listening on UDP {LISTEN_
PORT}, forwarding to UDP {FORWARD_PORT}")
while True:
data, addr = sock.recvfrom(1024)
if len(data) < 32:
print(f" Received too short packet from {addr}, ignored.")
continue
try:
decrypted = decrypt_data_chacha(data)
if len(decrypted) < 8:
print(f" Decrypted packet too short: {decrypted.hex()}")
continue
forward_socket.sendto(decrypted, ('127.0.0.1', FORWARD_
PORT))
print(f" Forwarded decrypted MAVLink message from {addr}
({len(decrypted)} bytes)")
except Exception as e:
print(f" Decryption failed from {addr}: {e}")
```

(E) MavlinkSender.py

```
import socket
import random
import time
```

```
from pymavlink.dialects.v20 import common as mavlink2
from AES import encrypt_data
from CHACHA20 import encrypt_data_chacha
encrypt_method=None
def random_send_mavlink(mav, udp_socket, target_addr):
message_type = random.choice(["heartbeat", "attitude",
"global_position_int", "command_long"])
if message_type == "heartbeat":
msg = mavlink2.MAVLink_heartbeat_message(
type=random.randint(0, 30),
autopilot=random.randint(0, 20),
base_mode=random.randint(0, 255),
custom_mode=random.randint(0, 1000),
system_status=random.randint(0, 10),
mavlink_version=3
)
elif message_type == "attitude":
msg = mavlink2.MAVLink_attitude_message(
time_boot_ms=int(time.time()*1000) % 0xFFFFFFFF,
roll=random.uniform(-3.14, 3.14),
pitch=random.uniform(-1.57, 1.57),
yaw=random.uniform(-3.14, 3.14),
rollspeed=random.uniform(-0.5, 0.5),
pitchspeed=random.uniform(-0.5, 0.5),
yawspeed=random.uniform(-0.5, 0.5)
)
elif message_type == "global_position_int":
msg = mavlink2.MAVLink_global_position_int_message(
time_boot_ms=int(time.time()*1000) % 0xFFFFFFFF,
lat=random.randint(-900000000, 900000000),
lon=random.randint(-1800000000, 1800000000),
alt=random.randint(0, 1000000),
relative_alt=random.randint(0, 1000000),
vx=random.randint(-1000, 1000),
vy=random.randint(-1000, 1000),
vz=random.randint(-1000, 1000),
hdg=random.randint(0, 35999)
)
elif message_type == "command_long":
msg = mavlink2.MAVLink_command_long_message(
```

```
target_system=1,
target_component=1,
command=random.choice([21, 22, 176, 400]),
confirmation=0,
param1=random.uniform(0, 100),
param2=random.uniform(0, 100),
param3=random.uniform(0, 100),
param4=random.uniform(0, 100),
param5=random.uniform(-90, 90),
param6=random.uniform(-180, 180),
param7=random.uniform(0, 100)
)
else:
print(" Unknown message type.")
return
# pack the massage into binary
binary_data = msg.pack(mav)
#encryption
encrypted_data = encrypt_method(binary_data)
#
udp_socket.sendto(encrypted_data, target_addr)
print(f" Sent {message_type} to {target_addr}")
def main():
udp_socket = socket.socket(socket.AF_INET, socket.SOCK_
DGRAM)
target_ip = '127.0.0.1'
target_port = 9000
target_addr = (target_ip, target_port)
mav = mavlink2.MAVLink(udp_socket)
print(f"□ Start sending random MAVLink messages to {target_
ip}:{target_port}")
try:
while True:
random_send_mavlink(mav, udp_socket, target_addr)
time.sleep(1) # one per second
except KeyboardInterrupt:
print("\n Stopped by user.")
finally:
udp_socket.close()
if __name__ == "__main__":
```

```
choice = input("choose chacha20 or AES: ").strip().lower()
if choice == "chacha20":
encrypt_method = encrypt_data_chacha
elif choice == "aes":
encrypt_method = encrypt_data
else:
print(" Invalid choice. Defaulting to AES.")
encrypt_method = encrypt_data
main()
```

(F)MavlinkReceiver.py

```
import socket
import time
import json
from pymavlink import mavutil
from writeCSV import *
def udp_listener(listen_ip='0.0.0.0', listen_port=14550):
print(f" Listening for MAVLink messages on {listen_ip}:{lis-
ten_port}...")
connection = mavutil.mavlink_connection(f'udp:{listen_ip-
}:{listen_port}', source_system=255)
last_seq_map = {} # system_id -> last_seq
total_packets = 0
total_lost = 0
try:
while True:
msg = connection.recv_msg()
if msg:
total_packets += 1
print("\n Received a MAVLink message:")
print(f" Type: {msg.get_type()}")
print(f" Message ID: {msg.get_msgId()}")
print(f" System ID: {msg.get_srcSystem()}")
print(f" Component ID: {msg.get_srcComponent()}")
print(f" Seq: {msg.get_seq()}")
raw_binary = msg.get_msgbuf()
print(f" Raw Binary: {raw_binary.hex()}")
decoded_dict = msg.to_dict()
print(" Decoded Content:")
print(json.dumps(decoded_dict, indent=2, default=lambda o:
o.hex() if isinstance(o, (bytes, bytearray)) else str(o)))
```

```
write_msg_to_csv(msg, 'mavlink_log.csv')
system_id = msg.get_srcSystem()
seq = msg.get_seq()
last_seq = last_seq_map.get(system_id)
if last_seq is not None:
expected_seq = (last_seq + 1) % 256
if seq != expected_seq:
lost = (seq - expected_seq) % 256
total_lost += lost
print(f" Detected {lost} packet(s) lost! (System ID: {sys-
tem_id}, Last seq: {last_seq}, Current seq: {seq})")
else:
print(f" First packet received from System ID {system_id}")
last_seq_map[system_id] = seq
loss_rate = (total_lost / (total_packets + total_lost)) *
100
print(f" Packets: {total_packets}, Lost: {total_lost}, Loss
rate: {loss_rate:.2f}%")
except KeyboardInterrupt:
print("\nStopped by user.")
if __name__ == "__main__":
udp_listener(listen_port=14550)
(G) writeCSV.py
import csv
import os
def write_msg_to_csv(msg, filename):
data = msg.to_dict()
data['msg_type'] = msg.get_type()
data['system_id'] = msg.get_srcSystem()
data['component_id'] = msg.get_srcComponent()
data['seq'] = msg.get_seq()
file_exists = os.path.isfile(filename)
with open(filename, mode='a', newline='') as f:
writer = csv.DictWriter(f, fieldnames=data.keys())
if not file_exists:
writer.writeheader()
writer.writerow(data)
```

Appendix B: Autoencoder Training Output

The code of the Autoencoder Implementation will be in the Autoencoder.zip or https://colab.research.google.com/drive/1P7WbT0K57bLFcyLWRBlbho rC7aFqhZtT?usp=sharing

Step 1:Load and Preview Data

```
import pandas as pd
data = pd.read_csv("mavlink_log.csv", on_bad_lines='skip')
print("Data preview:")
print(data.head())
print("\n Columns available:", data.columns.tolist())
```

Step 2:Data Cleaning and Feature Selection

```
import numpy as np
from sklearn.preprocessing import MinMaxScaler
```

1. Add missing columns if not present

```
if 'label' not in data.columns:
data['label'] = 0 # Assume all normal if no label
if 'anomaly_type' not in data.columns:
data['anomaly_type'] = 'unknown'
```

2. Select numerical features to use (adjust to match your CSV)

```
features = ['alt', 'vx', 'vy'] # ← Change here if needed
```

3. Drop rows with non-numeric or missing values

```
X_raw = data[features].apply(pd.to_numeric, errors='coerce')
# convert to NaN if not numeric
X_clean = X_raw.dropna() # remove rows with NaN
```

4. Sync labels and metadata

```
data_clean = data.loc[X_clean.index] # keep labels for clean
data
```

5. Normalize the features

```
scaler = MinMaxScaler()
X_scaled = scaler.fit_transform(X_clean)
```

6. Prepare training and testing sets

```
X_train = X_scaled[data_clean['label'] == 0]
X_test = X_scaled
y_test = data_clean['label'].values
print("Total number of data points:", len(data))
print("After converting to numeric values:", len(X_raw))
print("After removing missing values:", len(X_clean))
print("Preprocessing complete!")
print("X_train shape:", X_train.shape)
print("X_test shape:", X_test.shape)
```

Step 3:Autoencoder Model Building and Training

```
from tensorflow.keras.models import Model
from tensorflow.keras.layers import Input, Dense
from tensorflow.keras.optimizers import Adam
```

1. Define input dimension based on selected features

```
input_dim = X_train.shape[1]
```

2. Build Autoencoder architecture

```
input_layer = Input(shape=(input_dim,))
encoded = Dense(4, activation="relu")(input_layer)
encoded = Dense(2, activation="relu")(encoded)
decoded = Dense(4, activation="relu")(encoded)
output_layer = Dense(input_dim, activation="sigmoid")(decod-
ed)
autoencoder = Model(inputs=input_layer, outputs=output_lay-
er)
autoencoder.compile(optimizer=Adam(learning_rate=0.001),
loss='mse')
```

3. Train Autoencoder using only normal data

```
history = autoencoder.fit(
X_train, X_train,
epochs=50,
batch_size=32,
validation_split=0.1,
shuffle=True,
verbose=1
```

Step 4: Reconstruction Error and Thresholding

```
import numpy as np
```

1. Use Autoencoder to reconstruct test data

```
reconstructions = autoencoder.predict(X_test)
```

2. Calculate reconstruction error (MSE for each sample)

```
mse = np.mean(np.square(X_test - reconstructions), axis=1)
```

3. Determine anomaly threshold

```
# Method: mean + 3 * std (you can also use np.percentile if preferred)
threshold = mse.mean() + 3 * mse.std()
print("Threshold for anomaly detection:", round(threshold, 6))
```

4. Flag samples as anomalies if MSE > threshold

```
predictions = (mse > threshold).astype(int)
```

5. Add results to DataFrame

```
data_clean = data_clean.copy() # avoid modifying original
data_clean['mse'] = mse
data_clean['predicted_anomaly'] = predictions
```

6. Show how many anomalies were detected

```
n_anomalies = sum(predictions)
print(f"Detected {n_anomalies} anomalies out of {len(predictions)} samples")
```

Step 5: Anomaly Detection and Evaluation

```
from sklearn.metrics import confusion_matrix, classifica-
tion_report
import matplotlib.pyplot as plt
import seaborn as sns
```

1. Retrieve actual and predicted labels

```
y_true = data_clean['label'].values # Ground truth labels
y_pred = data_clean['predicted_anomaly'] # Labels predicted
by the Autoencoder
```

2. Generate and display the Confusion Matrix

```
cm = confusion_matrix(y_true, y_pred)
plt.figure(figsize=(5, 4))
sns.heatmap(cm, annot=True, fmt='d', cmap='Blues',
xticklabels=['Normal', 'Anomaly'],
yticklabels=['Normal', 'Anomaly'])
plt.title('Confusion Matrix')
plt.xlabel('Predicted')
plt.ylabel('Actual')
plt.show()
```

3. Display Classification Report

```
print("Classification Report:")
# Automatically infer labels from the data
print(classification_report(y_true, y_pred))
```

Step 6: Visualize Reconstruction Error by Anomaly Type

```
import matplotlib.pyplot as plt
import pandas as pd
import numpy as np
# Compute reconstruction error (difference between original and predicted data)
reconstructions = autoencoder.predict(X_test)
mse = np.mean(np.square(X_test - reconstructions), axis=1)
# Create a DataFrame including anomaly type
mse_df = pd.DataFrame({
'mse': mse,
'anomaly_type': data_clean['anomaly_type'].values # 'data_clean' is the preprocessed dataset
})
# Define anomaly threshold (e.g., 95th percentile)
threshold = np.percentile(mse, 95)
# Plot reconstruction error by anomaly type with different colors
plt.figure(figsize=(10, 6))
for anomaly_type in mse_df['anomaly_type'].unique():
subset = mse_df[mse_df['anomaly_type'] == anomaly_type]
plt.scatter(subset.index, subset['mse'], label=anomaly_type)
# Display threshold as a red dashed line
plt.axhline(threshold, color='red', linestyle='--', label='Threshold')
plt.title("Reconstruction Error by Anomaly Type")
plt.xlabel("Sample Index")
plt.ylabel("MSE (Reconstruction Error)")
plt.legend()
plt.grid(True)
plt.tight_layout()
plt.show()
```

Step 7: Anomaly Pattern Summary Table

```
# Define short descriptions for each anomaly type
```

```
anomaly_descriptions = {
'normal': 'Normal flight pattern',
'speed_abnormal': 'Abnormally high speed (e.g., >50 m/s)',
'battery_drop': 'Critically low battery (<10%)',
'altitude_abnormal': 'Extremely high altitude (>10,000 m)',
'unknown': 'Unspecified or mixed anomaly'
}
# Calculate the average reconstruction error (MSE) for each
anomaly type
mse_summary = mse_df.groupby('anomaly_type')['mse'].mean().
reset_index()
mse_summary = mse_summary.sort_values(by='mse', ascending=-
False)
# Define a function to classify detection ease based on av-
erage MSE
def detection_ease(mse_value):
if mse_value > 0.15:
return "Easy to detect"
elif mse_value > 0.08:
return "Moderate"
else:
return "Hard to detect"
# Build a summary list with anomaly type, description, aver-
age MSE, and detectability
pattern_list = []
for idx, row in mse_summary.iterrows():
anomaly_type = row['anomaly_type']
avg_mse = row['mse']
pattern_list.append({
'Anomaly Type': anomaly_type,
'Main Anomalous Feature': anomaly_descriptions.get(anomaly_
type, 'Unknown'),
'Average MSE': round(avg_mse, 4),
'Detectability': detection_ease(avg_mse)
})
# Convert the summary to a DataFrame and display it as a
Markdown table
pattern_df = pd.DataFrame(pattern_list)
print("Anomaly Pattern Summary Table")
```

```
print(pattern_df.to_markdown(index=False)) # Display as a
formatted table (Markdown style)
```

Step 8: Export Anomaly Pattern List as Markdown

```
# Convert the anomaly pattern summary to Markdown format
markdown_table = pattern_df.to_markdown(index=False)
print("Anomaly Pattern Summary (Markdown Format):")
print(markdown_table)
```

Chapter 3
Green Maritime Technologies:
The Role of Automation in Reducing Emissions and Enhancing Sustainability

Froilan Delute Mobo
https://orcid.org/0000-0002-4531-8106
Philippine Merchant Marine Academy, Philippines

Katarzyna Miłek
https://orcid.org/0000-0003-4603-4604
Akademia Nauk Stosowanych in Wrocław, Poland

ABSTRACT

Due to the negative environmental impact of international trade through the maritime sector, this industry is gaining disproportionate attention. The development of 'green' maritime technologies is a necessary response to global pressures that seek to reduce greenhouse gas emissions and increase sustainability. Automation through artificial intelligence, including route optimization, autonomous navigation, and smart port systems, is a groundbreaking approach to reducing fuel and emission dependencies. This chapter analyzes the shifts in maritime operations through the lens of the integration of automation and artificial intelligence in 'green' technologies, the regulatory and real-world aspects of compliance, the new environmental benefits, challenges, and opportunities in the quest for 'sustainable' maritime operations.

DOI: 10.4018/979-8-3373-3546-9.ch003

1. INTRODUCTION

The global economy benefits greatly from the ease with which the maritime industry transports raw materials and goods over international waters. According to the United Nations Conference on Trade and Development (UNCTAD), international shipping is an important section of the global supply chain and handles over 80% and 70% of the world trade volume and world trade value respectively (UNCTAD, 2023). The shipping industry facilitates globalization and connects markets. The shipping industry moves everything from energy goods and industrial equipment to food and consumer goods.

The economic benefits may not offset environmental costs. Most pollutants and greenhouse gases (GHGs) are produced during waterborne trade. The International Maritime Organization (IMO) estimated the percentage of global anthropogenic GHG emissions attributed to the shipping industry to be 2.89% (IMO, 2020) and predicts this will increase if GHG mitigation efforts are not strengthened. Shipping emissions contribute to global warming, but more immediately, and dangerously, they affect the health of people and the communities surrounding the ports and coasts of active shipping routes because they can emit the pollutants of PM, CO2, NOx, SOx, and PM to flow freely and uncontrollably into the atmosphere (Eyring et al., 2010; Corbett et al., 2007).

The environmental consequences of shipping have caught the attention of policymakers, environmentalists, and global institutions. Such actors have deployed ambitious regulatory structures and strategic plans to respond to the pressure on the shipping industry to implement sustainable initiatives and reduce emissions. The IMO's Initial GHG Strategy, implemented in 2018, aims to accomplish total decarbonization of international shipping and an annual gross reduction of GHG emissions (by 50% compared to 2008 levels) by 2050 (IMO, 2018). Achieving those objectives will necessitate radical transformations in ship design, fuel consumption, port operations, and logistical practices.

The development and use of green technologies in the shipping industry can provide maritime decarbonization. Some of the innovations include energy-efficient ship designs, wind and solar-assisted propulsion, and alternative fuels such as LNG, hydrogen, and ammonia. These technologies are designed to reduce the ecological effects of maritime activities. Although these innovations will address some of the fuel-related and structural elements of maritime sustainability, automation, the other structural component, is far more advanced and is an essential element in the industry's move toward more sustainable operations.

Automation in the maritime industry uses new technologies, such as artificial intelligence (AI), the Internet of Things (IoT), machine learning, big data, analytics, robotics, and satellite communications, to execute tasks that have historically been performed by human personnel. Automation improves situational awareness, streamlines cargo handling, optimizes vessel performance, and enhances the scheduling of maintenance. Automation can be applied to navigation of the vessel, control of the vessel's propulsion, monitoring of emissions, predictive maintenance, cargo management, and even a multitude of port activity tasks (Zhao et al., 2019; Psaraftis, 2019). For instance, by using real-time weather and oceanographic data, autonomous navigation systems can be designed to fuel-efficiently reroute and avoid inclement weather, high-resistance currents, and fuel-inefficient waterways. AI systems also control the vessel's engine loads, management of auxiliary systems, and hybrid propulsion modes to optimize fuel use and reduce emissions (Levander, 2017). These systems improve operational reliability and significantly reduce human error, a leading cause of maritime accidents, while also enhancing compliance with regulations (Kaber & Endsley, 2004).

The implementation of smart logistics and digital tools is vital for sustainable automation in port ecosystems. Automated steel-and-cable cranes with electric tugs and real-time data cut emissions and increase the speed and efficiency of cargo processing. Moreover, port call optimization tools allow transit vessels to adjust speeds and minimize the time spent idling at anchor, which helps to decrease fuel waste and congestion in the port (Winnes et al., 2015). To mesh with these systems, the largest ports are adopting "Just-in-Time" (JIT) arrival practices to coordinate ship and port readiness (UNCTAD, 2021).

The construction of Maritime Autonomous Surface Ships (MASS) also depends on the fusion of automation and green technologies in the maritime industry. Such vessels constitute a major shift in the industry as they are able to operate with negligible human supervision. MASS can be constructed with lighter and more fuel-efficient designs as the elimination of crew accommodations and optimized variable operational settings become the new operational standards (BIMCO, 2021; DNV, 2021). The first fully electric, self-sufficient, and self-operated container ship Yara Birkeland, launched in Norway, provides a perfect example of the fusion of automation, sustainability, and the transformation of the shipping industry (DNV, 2021). There are, however, challenges to automation in the industry along with insufficient potential. Issues of high capital costs, lack of synergy in procedures, cybersecurity, and vague regulatory environments all present significant transition challenges. The move to automated systems can disrupt legacy maritime labor systems and present questions of job loss and retraining (Psaraftis, 2019). Still, the potential of automation to meet environmental and economic ends makes it a viable candidate to progress towards global sustainability.This chapter considers how automation could change how the green goals of the maritime sector are pursued. The chapter looks into how automated technologies can facilitate the implementation of the IMO decarbonization strategy, help with fuel efficiency, and reduce emissions.

By examining advancements in the integration of automation with renewable energy and green fuels, and the current uses of automation within ports and on ships, this chapter documents the dynamics of maritime automation. To encourage their widespread adoption and to maximize the potential benefits to the environment, this paper analyzes the challenges and policy gaps that need to be addressed.

2. THE MARITIME ENVIRONMENTAL FOOTPRINT AND REGULATORY DRIVERS

Thanks to maritime transport, fundamental international trade continues to thrive. It enables the movement of bulk cargo, consumer goods, energy resources, and other essential tradeable items across continents. However, even with the economic value of maritime transport, it still contributes to the decline of the environment, particularly through the emission of greenhouse gases, poisonous gases, and other air pollutants. While the engines, auxiliary power units, and other onboard systems of ships combust fossil fuels, it results in the emission of gases of great concern. The consequences of these emissions go beyond the air and include the ocean biodiversity, the health of the ocean, and the global climate systems.

2.1 Profile of Emissions

The pollutant emissions which consist of carbon dioxide (CO_2), sulfur oxides (SO_X), nitrogen oxides (NO_X), and particulate matter (PM), all serve as primary indicators with regards to environment pollution the maritime sector. The primary source of these pollutants is the combustion of marine fuel used in the shipping industry, especially burning heavy fuel oils (HFO) which is the most polluting fuel (Brynolf et al., 2014). HFO is subject to widespread use due to its low costs and high energy density. Emissions from shipping activities and vessels account for almost 2.89% of global anthropogenic GHG emissions. From the climate change perspective, the most concerning emissions are the CO_2 emissions, which are the by-products of shipping activities (IMO, 2020). Depending on its operational profile and fuel type, a single large container vessel can emit over 200 metric tons of CO_2 daily (Smith et al., 2016). Since, with the anticipated growth of global trade, the size of these vessels is likely to grow. Hence, there is a need to reduce the emissions otherwise the projected emissions will be larger.

Burning sulfur-rich HFO is the main source of SO_X emissions. Acid rain, which can harm ecosystems and cause ocean acidification, is largely caused by these emissions. According to Smith et al. (2016), the 15 biggest container ships in the world are thought to emit as much sulfur as all 760 million cars combined. Sulfur dioxide negatively impacts human health and air quality, especially in port cities and along important shipping lanes, even though it does not directly cause global warming. Smog and ground-level ozone are primarily caused by NO_X emissions, which are produced during high-temperature combustion. They contribute to the eutrophication of marine ecosystems and worsen respiratory conditions like bronchitis and asthma. According to Corbett et al. (2007), particulate matter (PM), particularly PM2.5, can enter the human bloodstream and lungs deeply, leading to respiratory and cardiovascular diseases. Black carbon (BC), a short-lived climate pollutant with a high potential for global warming, is also released during the combustion process. In addition to increasing the heat absorption of atmospheric particles, black carbon from ships also contributes to Arctic warming by accumulating on ice and snow (Bond et al., 2013). BC has been identified by the IMO as a concern, especially in polar waters where its effects are more pronounced.

The distribution of emissions in the world's oceans is not consistent. They are primarily found near crowded ports like Shanghai, Singapore, Rotterdam, and Los Angeles, as well as in important maritime routes and port areas like the English Channel, the South China Sea, and the Gulf of Mexico. These regional emissions put the health of the local populace at risk and worsen air pollution in coastal regions (Eyring et al., 2010). Furthermore, because maritime transportation is an international activity, it is challenging to control and monitor shipping emissions. Convenience flags are frequently used by vessels, which makes it more difficult to enforce environmental laws. Furthermore, it is difficult to assign emissions accountability amongst stakeholders, such as shipowners, charterers, and cargo owners, due to the complexity of supply chains (Psaraftis, 2019). All things considered, the emissions profile of the maritime industry poses a variety of issues that go beyond climate change to include public health, marine ecology, and geopolitical governance.

2.2 Regulatory Drivers

Policies aimed at reducing the negative consequences of climate change in shipping are primarily the work of the International Maritime Organization (IMO) and are responding to the concern of the shipping industry about its environmental impact. Most of these policies are the consequence of the changes in the behavior of the shipping industry, going toward the adoption of new technologies. The shipping industry and air pollution regulations are primarily governed by the MARPOL Convention, especially Annex VI. The MARPOL Annex VI regulations include emission control of SO_X and NO_X, the prohibition of the intentional emissions of ozone-depleting substances, and energy efficiency standards. The most important MARPOL update occurred on January 1, 2020, when the global sulfur limit in marine fuels outside of emission control areas (ECAs) was decreased to 0.5% (by mass), which is significantly lower than the previous 3.5% limit (IMO, 2020). Moreover, the ECA limit is even more stringent, at 0.1% sulfur content. As a result of these new policies, many shipping companies adopted liquefied natural gas (LNG) as a bunker fuel, installed exhaust gas scrubbers, and/or used low-sulfur fuels.

In 2020, the Energy Efficiency Existing Ship Index (EEXI) and the Carbon Intensity Indicator (CII) were implemented around the globe for use and passed into law as regulations for use beginning 2023. These tools seek to optimize operational efficiency and reduce greenhouse gas emissions. The EEXI applies to existing vessels and, as the name implies, "energy efficiency" means assessing a ship's design and technical parameters, and for vessels that fall below the requisite efficiency standards, retrofits and design enhancements are mandated (IMO, 2020). Unlike EEXI, the CII examines the operational carbon intensity of the ship, specifically targeting the operational level of the ship. The CII determines the carbon dioxide emissions per transportation work (e.g., grams of CO2 per deadweight ton-mile) and ranks the vessels from A to E (best to worst) per "tax," and those that fall to E must provide a revision compliance plan to the regulations to avoid penalties (IMO, 2021). In addition to decreasing emissions, these tools are intended to provide market-based incentives. For example, vessels that score higher based on the CII are more likely to be chosen by charterers and clients, which can lead to increased revenue for ship owners who invest in automation and other green technologies.

The European Union and other regional players have begun some of their own initiatives as well. Under the EU Emissions Trading System (ETS), which expands to the EU maritime sector in 2024 (European Commission, 2023), shipping companies in EU waters will have to buy CO2 emission allowances. This will encourage further technological progress as the cost-benefit analysis of strategically decarbonizing shipping will be positively influenced. The more distant relevant strategy is the Chilean Government adopted the IMO's Initial GHG strategy in 2018, which aims for net zero emissions by the end of the century and a 50% reduction in shipping emissions from 2008 levels by 2050 (IMO, 2018). This strategy proposes energy efficient ship design, operational improvements, use of decarbonized ship fuels, and a series of economic and regulatory measures to be implemented in the shipping industry to achieve the IMO's 2020 GHG goals. The shift in the regulatory landscape for the shipping industry is now being focused on the incorporation of automation and digitization solutions for compliance. Automated emission control and emission compliance systems, AI based and autonomous cruise systems, smart maritime and autonomous systems for maintenance and repair provide real time tracking and control required for compliance to the radical new maritime emissions regulations.

3. AUTOMATION IN MARITIME TECHNOLOGIES

One of the most cutting-edge advancements in the field of automation in the maritime industry is the development of autonomous vessels, or Maritime Autonomous Surface Ships (MASS). The navigation, propulsion, collision avoidance, and other essential functions of these vessels are performed, at least in part, automatically, without much or any human involvement (Porathe et al., 2018). Autonomous vessels are equipped with integrated sensor systems consisting of radar, LiDAR, sonar, infrared cameras, and GPS, which are coupled with automatic Identification systems (AIS). These sensors provide a complete picture of the ship's surroundings. Onboard computers, equipped with risk assessment and machine learning algorithms, predict the behavior of proximate vessels and generate safe, efficient routes for passage. Supervisory and, if necessary, intervening control are available through remote operation. Autonomous vessels lower human error drastically, which is the cause of 75-96% of marine accidents (BIMCO, 2021). These vessels enhance safety by automating navigation and decision-making in intricate and perilous waterways. Furthermore, the elimination of crew quarters, galleys, and other life support systems increases design optimization within the vessels. Control systems with artificial intelligence and machine learning algorithms help with risk assessment and predict behavior. Consequently, there is a reduction in the ship's weight along with an increase in its cargo capacity. This, in turn, improves energy efficiency and decreases fuel use (Levander, 2017).

Decarbonization goals are also aided by autonomous vessels through real-time fuel optimization and use of alternative energy sources. To achieve complete operational efficiency while maintaining minimal emissions, hybrid autonomous vessels are able to automatically adjust battery and fossil fuel engine use depending on operational conditions (EMSA, 2022). The Yara Birkeland, launched in 2021 in Norway, is the most prominent real-world example of an autonomous vessel. Yara International and Kongsberg Maritime created the world's first fully electric, zero-emission autonomous container ship (DNV, 2021). The Yara Birkeland, which operates on a 7 MWh battery system, can fully eliminate 40,000 truck trips a year by making the Herøya and Brevik route, which has a cargo capacity of 120 TEU. It is remotely controlled and monitored from an onshore control center and has autonomous navigation, mooring, and loading systems. The Yara Birkeland serves as a model for the future of short-sea shipping by illustrating the profound synergies of automation and electrification. In addition to proving the technological viability of the system, the project also draws attention to the needed improvements in the infrastructure and the regulatory framework around such projects.

The Yara Birkeland requires specialized test sites, assured communication in real time, and dedicated legal frameworks concerning liability and compliance, which remain a work in progress nationally and internationally (IMO, 2021). Other pilot projects, such as Japan's Autonomous Urban Transport Project and the Mayflower Autonomous Ship, showcase the continued interest in this technology, which aims to decrease emission, enhance fuel efficiency, and reduce labor costs—aligning with the IMO GHG reduction strategies (Brynolf et al., 2022). The use of autonomous vessels raises ethical, legal, and technical concerns. Risks include system failures, cybersecurity gaps, accident liability, and the potential impact of autonomous vessels on the labor force. The need to craft robust policy frameworks for MASS certification, insurance, and global operational norms remains a priority for the maritime community. Autonomous ships, and MASS in general, epitomize the potential of maritime automation to transform the age of shipping to one that is more efficient, safer, and that reduces maritime pollution. Improved performance, innovative vessel designs, and a rethink of operational structures of these ships contribute to a radical reduction of emissions and a redefined global shipping industry.

Changing technology and legislation are likely to make the use of autonomous ships critical to the advancement of sustainable maritime transportation.

4. HOW AUTOMATION HELPS TO MINIMIZE EMISSIONS

The maritime shipping industry is under increased scrutiny from regulators and the public alike with respect to cost and its environmental footprint. This has made investment in automation a necessity to achieve enhanced energy efficiency and significant reductions in greenhouse gas (GHG) emissions. Automation technologies that have the capability to control a ship's operations in real-time and dynamically manage them have associated technologies to control and manage automated preventive maintenance, trim (fuel and emissions), and emissions compliance automation. This paper discusses the impact of automation on shipping in decarbonization and its four focal areas: route optimization; control of the engine and power systems; real-time emissions compliance; and automated predictive maintenance.

4.1 Emissions and Fuel Consumption on the Fly

Putting better methods to use in optimizing plans will certainly meet targets for both emissions and fuel use. One of the oldest and most conventional methods on the great seas has been to set a ‘route and timetable’ with little to no regard for the ever-changing and dynamic weather and sea patterns of winds, waves, and currents. Spending a lot of fuel and creating excessive emissions are the consequences of vessels avoiding severe weather or taking the longest, most circuitous, or easiest routes through congestion. Optimizing a route with AI solves this problem. It analyzes real-time weather and other environmental data, and a vessel's previously captured operating parameters data. In this way, it computes the most fuel-efficient and safe passage. An instance is the advanced Predictive Weather Analytics models of the Wärtsilä Fleet Operations Solution. It uses ML weather forecasting to improve the anticipated route performance of ongoing voyages (Wärtsilä 2022). It actively collects and analyzes data over the routes bound for geo areas forecasted (intended) to cover and avoid other zones with weather-band 'Adverse Weather' and areas with other types of severe ship resistance. In port, the saved fuel and time are subtracted from the time spent in emission reduction awaiting.

Ships using JIT principles in their coordination, where arrival at ports is timed, can reduce their speed to save on fuel even further.

As reported by the IMO (2020), adoption of slow steaming techniques is projected to reduce fuel consumption and greenhouse gas emissions by 30%. In addition, the optimization of greenhouse gas emissions and of CO2 ton-miles improves the overall efficiency of the voyage. Automated optimization of the route ensures that vessels sail under economic loading and the most favorable sea state, avoids unnecessary diversions, and complies with CII protocols (IMO, 2021), set by the IMO in 2023.

4.2 Power and Engine Control.

Another important contributor to emission reductions is the automation of power and engine control systems. With the manual operation of a traditional marine engine, fuel injection, timing of ignition, and combustion engine load distribution almost always become inefficient. Modern Engine Monitoring and Control Systems (EMCS) optimize performance under varying operational conditions by automating the real-time control of engine parameters through algorithms and sensors. EMCS, for example, lower fuel waste and greenhouse gas emissions by maintaining full combustion through proper control of complete fuel combustion at optimal conditions of post-combustion air, reclaimed exhaust gas, and the respective engine parameters. Minimization of Auxiliary Engine and Generator Idle Operation and fuel consumption is achieved by indirect control EMCS systems, which allow for the capture and controllable real-time adjustment of onboard energy supply demand (Corbett & Winebrake, 2008).

Many ships nowadays have hybrid propulsion, which addition to the standard diesel propulsion systems, also integrates internal combustion engines and battery-based propulsion systems. These systems, depending on the operational requirements, are able to change power sources automatically. For instance, the ship can operate on battery power only and at no point during the entire port maneuver, and for short distances above the water surface, thereby eliminating emissions. Optimized charge/discharge cycles and the use of electricity only when absolutely necessary by onboard diesel engines are primary-fueled managed systems that also meet the goals of EMSA(2022). Such hybrid designs are critical in regions with stringent emissions regulations, such as Emission Control Areas (ECAs). Electric energy stored during port maneuvers enables the vessel to meet SOx and NOx emission compliance without expensive fuel substitutes or exhaust scrubbers. Moreover, the automatic control with the lowest level of the negative side maintenance of oil changes, spare parts manufacture, engine repairs, circular supply central system repairs, and prolonged engine lifetime fuel savings. These maintenance actions happen with increased frequency due to the adverse repercussions that the engine control system inflicts on the overall sturdiness of the propulsion system.

4.3 Tracking and Reporting Emissions*

Tracking and reporting of emissions of international stature, like MARPOL Annex VI, must be conducted methodically and quantitatively. In the past, this process was achieved through a cumbersome approach of measuring and consulting the consumption logs of fuel over a set time period, and reporting on them. These processes continue to be plagued with inaccuracies, missed deadlines, and reporting discrepancies. In a controlled environment, emissions are computed and tracked in real time through a sophisticated amalgamation of autonomous systems and data from onboard sensors, data loggers, and cloud systems. For example, ABB's Marine Advisory System leverages sensors for real-time data on CO, CO2, NOx, SOx, and particulate matter and tracks operational behavior for computing suggestions (ABB 2021). Emission systems ensure consensus along with regulatory operational adjustments by setting feedback control emissions limits and recommending changes in auxiliary engines and vessel speed.

It is necessary to have emissions monitoring systems that enable tracking for real-time and automated emissions recording due to their necessity concerning the compliance scope and the accountability and transparency need for the monitoring systems especially due to the expansion of the European Union Emissions Trading System (EU ETS) to include the maritime sector. Companies that provide shipping services within the EU waters will need to purchase carbon credits to cover their emissions starting 2024 (2023 EU Commission). In addition, real-time reporting systems that include blockchain technology for emissions monitoring and reporting guarantee that emission monitoring systems are reliable, blockchain emission reporting systems, and provide automated, real-time, independent moratoria that block access to certain emissions data on demand to clients, government authorities, and moderators of the emission system in the ports. This enhanced transparency will foster stronger environmental stewardship and deepen the trust of the stakeholders.

In terms of cutting down on emissions, predictive maintenance might not be the first thing to come to the mind of an engineer, but it is still crucial. Predictive maintenance involves a system of scheduling within the repairable time frame, or waiting for the system to break down. The waiting approach leads to increased downtime, increased fuel consumption, and engine downgrades, which is the opposite of the ideal outcome we are pushing for. Predictive maintenance determines the functions of structures by observing real-time data and using AI and machine learning to correlate the data. Complex systems address deviations, preemptive failures, and abnormal diagnostics, and it used to function seamlessly. The system diagnoses unusual behavior to predict irregularities, downtimes, and breakdowns of the system. They detect wear and predict failures urgently to ensure continued functioning (Zhou et al, 2020). Predictive maintenance has lower emissions, fewer worn-out parts, and lower emergency spare parts, which are usually transported by air. Thus, the emissions from the engines are further accentuated.

In addition, integrating machine learning with digital twins, which are virtual models of a ship's systems, enables predictive maintenance systems to analyze potential breakdowns and recommend the best fixes. This type of simulation and modeling improves decision formulation, spare parts inventories, and ensures timely actions are taken on the ship, saving assets and time after the repairs (Lu et al, 2021).

5. INTEGRATION WITH GREEN TECHNOLOGIES

5.1 Battery Storage and Electrification

Direct emissions are eliminated by electric propulsion, which is controlled by automated systems. With the aid of AI-based energy management systems, battery-electric vessels are being used more frequently in short-sea shipping and ferries (EMSA, 2022).

5.2 Propulsion by Wind

Automation helps to optimize the use of renewable energy by adjusting rotors or wind sails. For example, to help with propulsion, the EnergySail system automatically aligns with the direction of the wind (IWSA, 2021).

5.3 Fuel Cells and Other Alternative Fuels

Complex alternative fuel systems, such as those involving ammonia, hydrogen, and liquefied natural gas (LNG), require automation to be managed. By ensuring safe fuel handling, combustion, and energy distribution, smart control systems maximize emission reductions (Brynolf et al., 2022).

6. SMART PORTS AND DIGITALIZATION

6.1 Optimization of Port Calls

IoT, automation, and data analytics are used by smart ports to control vessel arrival and departure, cutting down on idle emissions and waiting times. Port-ship communication systems facilitate just-in-time (JIT) arrivals, which allow ships to slow down and arrive precisely when a berth becomes available (UNCTAD, 2021).

6.2 Autonomous Cargo Management

Automated cranes and AGVs (Automated Guided Vehicles) in ports powered by electric systems reduce emissions and energy use compared to diesel-driven equipment. In order to achieve sustainability goals, ports such as Singapore and Rotterdam are setting the standard for automation and digitization (Port of Rotterdam, 2023; MPA Singapore, 2023).

7. CHALLENGES AND BARRIERS

7.1 Economic and Technological Obstacles

Widespread adoption is hampered by high capital costs, complicated system integration, and restricted retrofitting capabilities. Smaller shipping firms frequently don't have the funds to make a significant automation investment (Psaraftis, 2019).

7.2 Cybersecurity and Safety

Cyber threats and software malfunctions are among the new risks that automation brings. Regular software audits and a secure system architecture are essential for preventing intrusions into vessel navigation and control systems (BIMCO, 2021).

7.3 Inadequate Regulation

Regulation of automated and self-governing vessels is still developing globally. Although the IMO has started scoping exercises, uniform standards for operations, certification, and liability are still being developed (IMO, 2021).

8. CASE STUDIES

8.1 Norway's Yara Birkeland

By eliminating the need for 40,000 truck trips, this self-sufficient electric ship lowers CO_2 emissions by 678 tonnes yearly. It exhibits end-to-end automation in green shipping and is outfitted with automated mooring, navigation, and cargo loading systems (DNV, 2021).

8.2 Singapore's Port

The Maritime and Port Authority (MPA) of Singapore has made investments in automated berthing, AI-powered traffic control, and intelligent logistics hubs. According to MPA Singapore (2023), these initiatives have improved efficiency while lowering turnaround times and port emissions.

9. THE FUTURE OF AUTOMATED GREEN SHIPPING

The maritime sector is set to undergo revolutionary change as artificial intelligence, robotics, and renewable energy systems advance. Fully autonomous, emission-free ships might be the standard by 2050, especially for intraregional and short-haul shipping. Predictive analytics, operational effectiveness, and transparency will all be improved by blockchain, digital twins, and remote operation centers. In order to meet climate goals, these technologies must come together under the green maritime agenda.

10. CONCLUSION

Automation is a strong facilitator of maritime decarbonization, but it is not a cure-all. Automation facilitates the industry's transition to sustainability by improving efficiency, monitoring emissions, and integrating with renewable energy systems. Although there are still obstacles to overcome, particularly in the areas of cybersecurity and regulatory standardization, the future of shipping is clearly headed toward green automation.

REFERENCES

ABB. (2021). Marine Advisory System: Emissions monitoring for sustainable shipping. ABB Marine & Ports. https://new.abb.com/marine

ABB. (2021). ABB's Marine Advisory System: Driving sustainable shipping. https://global.abb

BIMCO. (2021). Cyber security guidelines for ships. https://www.bimco.org/cyber-security

Bond, T. C., Doherty, S. J., Fahey, D. W., Forster, P. M., Berntsen, T., DeAngelo, B. J., & Zender, C. S. (2013). Bounding the role of black carbon in the climate system: A scientific assessment. *Journal of Geophysical Research, D, Atmospheres*, *118*(11), 5380–5552. DOI: 10.1002/jgrd.50171

Brynolf, S., Andersson, K., & Ng, A. K. Y. (2014). Environmental assessment of marine fuels: Liquefied natural gas, liquefied biogas, methanol, and bio-methanol. *Journal of Cleaner Production*, *74*, 86–95. DOI: 10.1016/j.jclepro.2014.03.052

Brynolf, S., Fridell, E., & Andersson, K. (2022). Environmental assessment of hydrogen and ammonia as fuels for shipping. *Sustainable Energy & Fuels*, *6*(2), 300–318. DOI: 10.1039/D1SE01222F

Corbett, J. J., & Winebrake, J. J. (2008). The impacts of global maritime shipping on air pollution and climate change. *Environmental Science & Technology*, *42*(13), 4486–4493. DOI: 10.1021/es071686z

Corbett, J. J., Winebrake, J. J., Green, E. H., Kasibhatla, P., Eyring, V., & Lauer, A. (2007). Mortality from ship emissions: A global assessment. *Environmental Science & Technology*, *41*(24), 8512–8518. DOI: 10.1021/es071686z

DNV. (2021). Yara Birkeland: The world's first autonomous and electric container ship. https://www.dnv.com

EMSA. (2022). Study on electrical energy storage systems in maritime transport. European Maritime Safety Agency. https://www.emsa.europa.eu

European Commission. (2023). EU Emissions Trading System (EU ETS): Maritime sector inclusion. https://climate.ec.europa.eu

Eyring, V., Köhler, H. W., van Aardenne, J., & Lauer, A. (2010). Emissions from international shipping: 1. The last 50 years. *Journal of Geophysical Research, D, Atmospheres*, *115*(D12). Advance online publication. DOI: 10.1029/2009JD012396

IMO. (2018). Initial IMO Strategy on Reduction of GHG Emissions from Ships. https://www.imo.org

IMO. (2020). Fourth, IMO, GHG Study 2020. International Maritime Organization. https://www.imo.org

IMO. (2020). MARPOL Annex VI: Prevention of air pollution from ships. https://www.imo.org

IMO. (2021). Carbon Intensity Indicator (CII) and Energy Efficiency Existing Ship Index (EEXI). https://www.imo.org

IMO. (2021). Regulatory scoping exercise for the use of Maritime Autonomous Surface Ships (MASS). https://www.imo.org

IWSA. (2021). Wind propulsion technologies and systems. International Windship Association. https://wind-ship.org

Kaber, D. B., & Endsley, M. R. (2004). The effects of level of automation and adaptive automation on human performance, situation awareness, and workload in a dynamic control task. *Theoretical Issues in Ergonomics Science*, *5*(2), 113–153. DOI: 10.1080/1463922021000054335

Levander, O. (2017). Autonomous ships on the high seas. *IEEE Spectrum*, *54*(2), 26–31. DOI: 10.1109/MSPEC.2017.7849225

Lu, Y., Liu, C., & Xu, X. (2021). Digital twin-driven smart maintenance: Systems, technologies, and applications. *Journal of Manufacturing Systems*, *58*, 346–361. DOI: 10.1016/j.jmsy.2020.06.006

Singapore, M. P. A. (2023). Smart port initiatives. Maritime and Port Authority of Singapore. https://www.mpa.gov.sg

Porathe, T., Ehlers, S., & Jensen, M. (2018). Maritime Autonomous Surface Ships (MASS): State-of-the-art autonomous control. Proceedings of the 5th International Conference on Maritime Science and Technology. Port of Rotterdam. (2023). Digital and automated port operations. https://www.portofrotterdam.com

Psaraftis, H. N. (2019). Sustainable shipping: A cross-disciplinary view. Springer. Smith, T. W. P., Jalkanen, J. P., Anderson, B. A., Corbett, J. J., & Faber, J. (2016). CO_2 emissions from international shipping: Possible reduction targets and their associated pathways. IMO MEPC 68/INF.24. UNCTAD. (2021). Review of Maritime Transport 2021. United Nations Conference on Trade and Development. https://unctad.org

UNCTAD. (2023). Review of Maritime Transport 2023. https://unctad.org

Wärtsilä. (2022). Fleet operations solution for sustainable shipping. Wärtsilä Voyage. https://www.wartsila.com

Winnes, H., Styhre, L., & Fridell, E. (2015). Reducing GHG emissions from ships in port areas. *Research in Transportation Business & Management*, *17*, 73–82. DOI: 10.1016/j.rtbm.2015.10.008

Zhao, J., Shi, W., & Xu, M. (2019). The potential of autonomous ships in green shipping. *Ocean Engineering*, *189*, 106299. DOI: 10.1016/j.oceaneng.2019.106299

Zhou, Q., Liu, C., & Wu, Y. (2020). Predictive maintenance for ship propulsion systems using machine learning. *Journal of Marine Engineering & Technology*, *19*(5), 245–255. DOI: 10.1080/20464177.2020.1783002

Chapter 4
Machine Minds at Sea: A New Era of AI-Driven Maritime Intelligence

Alaina Thea Villanueva Concepcion
https://orcid.org/0009-0001-6874-6859
Bulacan State University, Philippines

ABSTRACT

Artificial Intelligence (AI) is changing how the marine business works in a big way. It is changing how environmental monitoring, security, cargo management, and navigation are done. AI-powered solutions are now used in the marine industry for things like predictive maintenance, route optimization, real-time data analysis, and managing ships on their own. In the past, these activities relied on manual decision-making and analog instruments. These "machine minds at sea" help keep the world's oceans safe, make things run more smoothly, cut down on mistakes made by people, and make things safer. AI also makes it possible to manage the environment in a proactive way by using sustainable routing, pollution monitoring, and emission tracking. Combining human knowledge with AI makes the global maritime ecosystem safer, more efficient, and more sustainable. This chapter talks about how AI is used, what its pros and cons are, and what it means for marine intelligence.

INTRODUCTION

The impact of the integration of AI nowadays in maritime transportation has a totally big impact on thc maritimc industry, by cultivating the efficiency of ships, safety, and risk management, according to Durlik, Miller, Kostecka, and Tuński (2024). The maritime industry is experiencing a standard shift in the big implementation of AI. As such radio communication and human decision-making use

DOI: 10.4018/979-8-3373-3546-9.ch004

predictive maintenance in maintaining ships and identifying route optimization in using advanced technologies applications. The transportation of goods in the maritime supply chain can also be enhanced with the advancement of AI, which can also detect threats detection within the maritime industry, and allow strategic planning. The challenge on the increasing intricacy of naval logistics and the globalized nature of trade, can change and address.

"Machine Minds" highlights the use of new technology which are changing the way how the data collection are being analyzed, and used during at sea. The Grand View Research (2024) says that the maritime artificial intelligence industry could grow a lot by 2030 because of advances and implementation in automation and use of data analytics. AI is very important, because it can identify weather in real time during sailing without any hassle, also by keeping track of underwater habitats, finding broken equipment, and predicting crashes or possible piracy. These technologies make things become more safer, more efficient, and less likely to make mistakes more aware of what's going on around them.

The AI implementation is driven by the growing implementation of AI in terms of safety, sustainability in the vessel industry, and security beyond efficiency gains. According to Grand View Research (2024), the maritime AI market is projected to experience swift growth by the year 2030 driven by automation and advanced analytics. Of course, challenges will still remain, including high legacy infrastructure incompatibility, the cost of implementation in the maritime industry, ethical issues, cybersecurity threats at sea, and workforce readiness gaps. According to the International Maritime Organization (2025), the AI managing risk and opportunities emphasizes the urgent need for a well-defined governance framework.

However, AI, in addition to marine infrastructure, is expensive and requires people and digital resources with specialized capabilities, with the help of training and seminars. According to the International Maritime Organization (2025), several barriers stand in the way of widespread adoption, including cybersecurity dangers, some ethical issues, regulatory loopholes, and worries over data privacy make widespread use much harder. We all know that a lot of older ships and port facilities haven't used AI systems before, which makes it hard for people to use them.

This chapter focuses on the current maritime challenges and offers a well-organized and coherent discussion. Firstly, measured data and real-world cases cause the discussion to lean more toward description rather than analysis. The AI's maritime industry looks magnified when you look at real-world results. Secondly, there are some repetitive ideas across sections, which focus on efficiency and effectiveness of AI, cybersecurity risks, and other regulatory issues, which weaken clarity and create unnecessary redundancy. Thirdly, the absence of a clear theoretical framework and conceptual framework makes it harder to follow the argument's logic, which could help to organize the discussion and strengthen the analytical complexity of the arguments. Fourthly, ethical and some human factors, such as crew training, workforce revision, and the psychological impact of automation, are stated briefly and deserve a more thorough examination to make the discussion holistic. Lastly, citations are inconsistent; in several points, information is presented without adequate scholarly backing, which diminishes the level of academic rigor required in this book chapter.

Objectives of this book chapter:

1. To aspect at just how AI is now being used in nautical operations, such as navigation, surveillance, logistics, and monitoring the environment.
2. To measure the benefits of AI-driven systems in improving decision-making, operational efficiency, and maritime safety.
3. To observe the obstacles and impediments in the integration of AI technologies, including legislative, technological, and ethical concerns, especially within conventional maritime infrastructures (Sameera Sulakshana, 2024).
4. To explore existing trends and future advancements in AI-driven marine intelligence and their ramifications for sustainability and global maritime security (Schill & Browning, 2015).

1. AI IN MARITIME OPERATIONS

The chapter demonstrates a stable attempt to cover the advantages, challenges, and strategic implications of AI-driven maritime intelligence, but there are areas of over-emphasis and under-emphasis. Although these are significant, they dominate much of the argument, leading to redundancy and overshadowing other equally critical dimensions. For instance, the points about route optimization, predictive maintenance, and condensed fuel ingesting are repeated in multiple sections, which weakens the impact of the analysis.

Before the digital uprising, marine intelligence differed a lot in people's construction decisions, radio communication, radar instruments, and paper maps for navigation. For a long time, these methods worked well for industry, but they weren't very good at responding to changes in real time (Pacific Data Integrators, n.d.). Navigators, ship captains, and port officials had to make effective decisions based on incomplete but important information, such as what they could see, hear, and read on radar and nautical charts.

Traditional maritime processes depend on human knowledge, manual error, and involvement-driven decision-making. Weather data, past information, and the knowledge of port managers and fleet operators helped with navigation, fleet coordination, and port operations. Sometimes the mistake made by humans delays the response time and misses some information or it is not accurate, AI approach will make it work well.

According to Mohammed & Ahmed (2025), AI and progressive digital systems have altered maritime intelligence that enables computerized decision-support systems that AI can provide real-time data procedure. The use of machine intelligence also works with human knowledge, skills, and their specialization. According to the International Maritime Organization (2025) it articulates that this integration is the start of a new era, where the use of AI radar systems helps in transforming the way crafts and vessels work by preparing them to be more heedful of their surroundings and be safer. Also, it lets owners identify quickly how operational data in maritime can be an eco-friendly environment, and it can also help logistics.

Key Components of Traditional Maritime Intelligence

Before the digital age, captains and officers, and the port authorities were some of the key people who made operational decisions based on their ability to observe, to use analog tools, and to follow established reporting systems. Maritime intelligence now depended on several personnel and shipboard machinery.

Manual Navigation and Charting

This method has worked well for generations, but it took a long time, and it was very easy to make mistakes, especially once visibility is limited or the weather is bad. To find out where a ship was at sea, traditional navigation and charting methods used paper nautical charts, compasses, sextants, and celestial navigation.

Radar and Sonar Systems

Vessels also use radar and sonar equipment, which involves talent and training to recognize raw data and make imperative choices. These technologies and knowledge were necessary for creating safer and more aware manual maritime processes.

- Radar (Radio Detection and Ranging) allowed ships to detect other vessels, landmasses, and obstacles at a distance, cultivating collision avoidance and overall safety.
- Sonar (Sound Navigation and Ranging) is sensitive for sensing underwater substances and mapping seabed's, mostly for naval operations, research, and fishing (Xia, W., et.al, 2022).

Radar (Radio Detection and Ranging) and Sonar (Sound Navigation and Ranging) are the same remote sensing technologies, though they use different types of waves to offer information. Sonars refer to sound waves, although radar refers to electromagnetic waves. In together systems, waves reflect off things to tell you in what way it is big, how far away, and how fast the target is moving. Radar is typically used to detect the surface and the air since electromagnetic radiation loses strength quickly in water. Sonar, on the other hand, works well in underwater environments, making it great for finding and navigating underwater (Xia, W., et.al, 2022).

Radio Communications

VHF radio plays a crucial role in the Global Maritime Distress and Safety System or also known as GMDSS. They offer weather information, navigational warnings, and emergency alerts, which are mostly vital for maritime communication between ships and shore posts.

Visual Observation and Watchkeeping

One of the tasks that is important in the maritime industry is that you need a maritime staff in constantly keep a close eye on the distance to look for dangers. Even with the use of modern navigation equipment, it is also needed to be alert at all times. Staff schedule scanning for potential hazards approaching vessels carefully which with the help of AI-driven technologies, improves situational awareness. However, maritime staff still had to look out for threats that radar and sonar couldn't find.

Intelligence Gathering Through Reports

Before, the maritime security and the maritime intelligence organizations relied only on reports from export ships, ports, and nautical patrols. AI helps process huge amounts of reports in maritime business it transforms the raw data into actionable intelligence which it can easily provide them with reports quickly. With this it increases risk management and safety within the maritime sector.

Emergence of AI Technologies in the Maritime Domain

Automatic data analysis and predictive maintenance are beginning to change marine operations because of the fast rise of artificial intelligence (AI) in the maritime industry (Hale, Long, Corns, & Shoberg, 2019). Navigation is now more efficient and dependable because of AI's capacity to interpret data in real-time and produce correct forecasts. This advancement is largely driven by the combined use of AI and machine learning technologies.

AI Applications in Radar, Sonar, and Vessel Monitoring

Artificial intelligence (AI) now works alongside human knowledge to improve safety and accuracy in maritime operations. It allows people to make faster and more reliable decisions at sea, which helps reduce risks and improve efficiency. Moreover, modern AI systems function similarly to sonar technology because they use sound waves to detect underwater obstacles and other possible dangers. Ships also make use of technologies like the Automatic Identification System (AIS), Global Positioning System (GPS), and sensors onboard to foretell where other ships will go, direct them along more secure routes, and alert workers to impending dangers like poor weather or accidents. Radar systems powered by AI can also identify, follow, and categorize ships and other objects in real-time. Therefore, these technologies greatly aid in safer navigation by reducing the likelihood of human error (Xia, Xing, & Guo, 2022). Additionally, AI has enhanced conventional technologies such as radar and sonar, which has improved the monitoring and guiding of ships. Vessels rely on these smart systems to keep them afloat and in the clear. In addition to keeping an eye out for marine life, they figured out the best paths for submarines and other boats. Using data in real-time, AI is making maritime transport smarter, safer, and more environmentally friendly.

Transformative Applications of Artificial Intelligence in Maritime Operations

These days, employing flexible frameworks that combine automated and human tasks to build operations safer, more efficient, and more strategic, and AI is used as a substitute for reactive systems. Artificial intelligence (AI) is being transformed into maritime operations that use machine learning and deep learning.

A. Predictive Maintenance

According to industry statistics, AI-powered predictive maintenance can cut expenses by as much as 10%. Predictive maintenance, powered by AI, makes equipment more reliable and available. Case studies show useful applications, such as the study of using real-time sensor data using methodologies like k-Nearest Neighbors (kNN) for problem recognition and disappointment reduction on vessels. AI models engaging autoencoders can also repeatedly oversee motors and detect misdeeds in real time, transforming from old preventive methods to sophisticated prediction techniques, (Almutairi & Khan, 2023).

B. Autonomous Navigation and Collision Avoidance

Autonomous navigation systems that use AI-powered sensors focus on fusion to combine the use of radar, LiDAR, AIS, GPS, and thermal imaging to make the vessel move easier and safely (Schill & Browning, 2015). Deep learning applications are becoming increasingly important for autonomous surface vehicles (ASVs). They make it possible for ASVs to do complicated tasks, including inspecting the coast, detecting the ocean, working together on missions, and directing search-and-rescue operations.

C. Anomaly Detection and Threat Monitoring

Artificial intelligence (AI) models are relatively good at discovery unusual behavior in vessels or ships, such as when AIS signals are changed. They can do this with about 99.9% accuracy by engaging artificial neural networks (ANNs) to keep an eye on speed, course, and timing data.

D. Operational Optimization and Sustainability

AI makes routing better, lowers the risks of bad weather, and uses less fuel. For example, DNV GL's research displays that AI-based course optimization can save up to 5% on fuel. AI also helps with load logistics, port operations, and keeping an eye on the environment. This makes results easier and has a far less impact on the environment. Global Plastic Watch and other programs use AI to keep an eye on plastic garbage in the ocean using satellite images and sound data.

E. Digital Twins and Reinforced Predictive Control

The idea of a numerical twin, which is a very precise virtual copy of a ship, is being used to expand situational awareness in ASVs. When used with strengthening knowledge and model predictive control, digital twins make it likely to make results based on scenarios and avoid collisions (Zhu, Wang, & Gao, 2025).

2. CORE COMPONENTS OF AI-DRIVEN MARITIME INTELLIGENCE

The core parts of AI-driven nautical intelligence contain the use of machine learning algorithms to associate massive amounts of old and real-time data. These systems suggest predictions, power tasks, and assist people in making improved choices in all maritime activities. Machine intelligence is one of the most vital technologies nowadays for sifting navigation, using surveillance, and application in actions at sea. By using AI models, those who work in the marine industry may forecast possible dangers, find the best routes for every vessel, focus on ship action or performance, and make it possible and make naval operations safer and more efficient at sea.

Key AI components include:

1. **Predictive Analytics -** AI algorithms examine historical and real-time data to forecast potential equipment failures, environmental risks, or operational tailbacks (Mohammed & Ahmed, 2025).
2. **Sensor Integration -** Radar, sonar, AIS, GPS, LiDAR, and eco-friendly sensors provide constant data streams for AI systems to focus on maritime conditions and vessel performance (Xia, Xing, & Guo, 2022; Schill & Browning, 2015).

All these portions work to form a strong AI-driven marine intelligence framework that improves operational resilience, situational perception, and long-term maritime follow-up.

2.1. Machine Learning Algorithms

Machine learning (ML) algorithms are charming and an important part of AI-driven marine intelligence. They make triangulation and investigation procedures improved. ML systems let vessels and naval authorities make choices quicker, cleverer, and more elastic than outdated approaches by looking at former datasets and real-time sensor inputs (Almutairi & Khan, 2023).

ML in Navigation

ML algorithms look at a lot of data, such as previous trips, ocean watercourses, recurrent weather movements, and records of port congestion. AI-based course optimization systems might consider things like wind speeds, wave heights, and measured areas. And, to find the best routes that combine safety, fuel effectiveness, and rapidity. This is changed from static charts. These algorithms are changing repeatedly while new information becomes available, such as updated weather forecasts and, or reports of incidents, come in.

ML in Surveillance

Machine learning (ML) approaches are vital for maritime surveillance. By looking at data from a lot of different places, such underwater sound sensors and radar systems, ML may find and identify acts that don't follow normal navigation patterns and strange behavior of vessels. This skill makes it easier to be aware of what's going on around you and safer at sea in general. By using high-resolution photos to find ships that don't send out Automatic Identification System (AIS) signals, ML integration in maritime operations not only makes things run more smoothly, but it also makes it easier to understand the data.

Benefits of ML in Maritime Operations

Machine learning (ML) is an important part of marine intelligence that uses AI. It has several useful uses, such as helping ship owners save money on running their businesses, making the ocean safer by anticipating when maintenance is needed, and safeguarding the maritime environment. ML is an integral aspect of AI-driven marine intelligence, which in turn improves the efficiency, predictability, and adaptability of the marine ecosystem.

2.2. Machine Learning in Maritime Surveillance

Machine learning (ML) is revolutionizing maritime surveillance by improving its ability to support regulatory compliance and various forms of marine domain awareness, including self-protection. Maritime surveillance systems use algorithms to process data streams from sources such underwater acoustic sensors, satellite images, Automatic Identification System (AIS) broadcasts, and radar returns (Schill & Browning, 2015). With the use of ML, they do not need to hire people anymore. These technologies work together to detect suspicious behavior on board, such as abrupt changes in direction or odd changes in speed, or the intentional deactivation of transponder patterns that indicate illicit fishing, smuggling, or other unethical activities.

Machine learning can help find cases of non-compliance, such when someone throws away trash illegally or enters a marine protected area. According to Mallam, Nazir, and Sharma (2020), ML makes it easier to keep an eye on commercial ships for possible dangers like piracy or strange approaches. ML is also helpful for keeping an eye on busy locations like the Panama Canal and the Strait of Malacca since it can find and report strange patterns in real time, which lets security forces respond quickly and effectively (European Maritime Safety Agency, 2022). These segments show how ML's innovative role in making the seas safer and protecting the environment has made operations more efficient.

2.3. Real-Time Data Analytics and Predictive Modeling

AI-driven maritime intelligence combines real-time data analysis and predictive modeling to help ships and maritime authorities stay one step ahead of dangers, make operations run more smoothly, and keep everyone safer. Maritime stakeholders can monitor ships, assess information as it comes in, and make reliable predictions about future developments by keeping an eye on the condition of ships, the state of their cargo, the weather, and traffic.

Real-Time Data Analytics

Real-time analytics means constantly gathering and analyzing data from many onboard and outside sources, such as GPS, radar, sonar, engine monitors, and weather sensors. This information is combined into a single operational picture that may be used to make decisions.

Key applications include:

1. **Voyage Monitoring -** Tracking vessel position, speed, and heading to ensure adherence to planned routes and detect deviations.
2. **Condition Monitoring -** Observing machinery health indicators (e.g., temperature, vibration, fuel usage) to detect potential failures early.
3. **Situational Awareness -** Combining AIS, radar, and satellite feeds to create a live picture of nearby weather, vessel traffic, and potential hazards.

Predictive Modeling

A dynamic decision-support system is made by putting real-time data and predictive models. AI algorithms and historical data, which is very important in forecasting, can guess what can happen in the future by predictive modeling services.

The integration of artificial intelligence and machine learning technologies provides multiple benefits to maritime operations, particularly in enhancing safety and ensuring environmental compliance. One of the most significant contributions is the enhancement of safety, as intelligent systems enable the early detection of hazards and allow for faster, data-driven response times, thereby reducing the likelihood of accidents and collisions (Haugen & Kristiansen, 2022). According to Rodseth & Burmeister (2015), operational efficiency is also strengthened with optimized routing and predictive maintenance minimizing downtime and ensuring smoother vessel operations

Lastly, AI and ML tools protect environmental compliance by predicting emissions that can monitor how vessels perform in making sure ships follow maritime operations with sustainability goals (IMO, 2021). These benefits demonstrate the transformative potential of AI-driven solutions in creating a safer travel during at sea, being more efficient, and environmentally responsible maritime environment.

The Port of Rotterdam utilizes AI-powered analytics to manage over 1.2 million ship arrivals and departures annually. This allows for creating more efficient disembarking schedules and reducing traffic congestion. Similarly, Maersk uses satellite data in real-time and weather models to adjust the transoceanic routes. This saves up to 5% on fuel costs per trip and makes the trips safer in rough seas.

2.4 Autonomous Vessel Control Systems

AI-powered autopilot systems are among the important applications of the new technologies in the maritime industry. These systems adjust in changing the sea conditions by their own, allowing ships to move, navigate and operate with minimal human assistance. Autonomous vessel control systems use AI, sensor fusion, and advanced control algorithms to make real-time choices about navigation and operations (IMO, 2021; Haugen & Kristiansen, 2022).

Autonomous vessel control systems are a big step toward ships that can navigate themselves completely. These systems make up an AI-powered control framework by putting together data from radar, sonar, the Global Positioning System (GPS), the Automatic Identification System (AIS), weather sensors, and high-resolution cameras. This architecture lets ships steer, avoid obstacles, control their speed, dock, and deal with unexpected situations, frequently without much or any help from people (Rodseth & Burmeister, 2015).

Most existing implementations only work at partial or conditional levels of autonomy. However, pioneering initiatives like the Mayflower Autonomous Ship (MAS) by ProMare and Yara Birkeland by Yara International show that fully autonomous maritime operations are possible. Expected benefits include better safety, more efficient operations, and fewer mistakes made by people. Still, technology has a lot of problems to deal with when it comes to rules, security, prices, and moral issues (DNV, 2021).

Defining Autonomous Vessel Control

An autonomous vessel control system uses AI to handle data from a variety of onboard and external sources. These are radar, sonar, GPS, AIS, weather monitoring systems, and technologies that create images. These systems can do the following by using machine learning models and rule-based navigation logic:

- Steer vessels along predefined itineraries
- Avoid obstacles and prevent collisions
- Adjust speed and direction in response to environmental conditions
- Execute docking and undocking protocols
- Respond to emergencies with minimal or no human intervention

Key Technologies Involved

To make autonomous ships work in control systems, numerous important technologies need to work together. Sensor fusion uses different sensors like radar, LiDAR, optical cameras, infrared imaging, and sonar to give a complete and accurate picture of the nautical environment. Machine learning algorithms help find impediments, choose the best routes for navigation, and guess where dangers might be. Control systems and actuators use AI-driven processes that allow steering and propulsion to happen automatically. This makes things more accurate and lessens the need for human input. Lastly, vessel-to-shore connectivity uses high-speed satellite communication to make it possible to monitor ships in real time and operate them from afar. This makes sure that the ship and onshore control centers are always in touch with each other.

Key operational advantages include:

1. **Enhanced Safety -** Faster responses to hazards reduce risks of collisions.
2. **Operational Efficiency -** Optimized routing and fuel utilization lower costs.
3. **Reduced Human Error -** Automation minimizes fatigue and oversight risks.
4. **Extended Operational Reach** - Enables vessel operation in hazardous environments.

Autonomous vessel control systems are a big change in how ships work. Even though businesses haven't widely adopted this technology yet, AI-driven decision-making, advanced sensor fusion, and strong communication networks make it a key part of the future of maritime transport. It promises safer, more efficient, and more sustainable global shipping (Haugen & Kristiansen, 2022).

2.5 Internet of Things (IoT) and Sensor Integration at Sea

The precision and dependability of data gathered from connected navigation and sensing technologies are what make AI-driven maritime systems work. The Internet of Things (IoT) can be used with maritime operations to offer real-time decision-making, predictive analytics, and continuous monitoring. Some of the most important parts are:

1. **Global Positioning System (GPS) -** Gives exact information on position, speed, and timing, which helps with navigation, route optimization, and ETA calculations. When combined with AI systems, GPS data makes route planning and operations more efficient. In autonomous vessels, GPS is used with inertial navigation systems to keep the ship on course when satellite signals are weak.
2. **Sonar (Sound Navigation and Ranging) -** Uses sound waves to find things in the water and make maps of the ocean floor. Some uses are safe travel in shallow or unknown seas, monitoring the health of marine ecosystems, and defense activities like finding submarines or drones. AI makes sonar better by sorting the items it finds and telling the difference between natural and man-made structures.
3. **Satellite Automatic Identification System (AIS) -** Keeps track of the identification, location, and operational status of ships all over the world. It helps with managing traffic, avoiding accidents, and enforcing rules. AIS works better with AI to help people understand what's going on in busy or dangerous sea lanes.
4. **Smart Sensors -** Sensors that work with the Internet of Things (IoT) keep an eye on mechanical systems, fuel efficiency, the weather, and pollution. These sensors send high-frequency data streams to AI models that help with predictive maintenance, safety, and following environmental rules.

Sensor fusion is the term for combining these technologies to get a complete picture of operations. GPS tells you where a ship is, sonar finds underwater dangers, AIS keeps track of traffic around a ship, and smart sensors provide you with information on the health of machinery. AI-driven control systems may use all these inputs to improve navigation, safety, and efficiency in modern maritime operations.

The Power of Data Integration

Each maritime technology gives useful information on its own, but when they work together, they give a complete and more accurate picture of how things are working. AI systems create a real-time and complete picture of the maritime environment by merging GPS positioning, sonar readings, AIS transmissions, and smart sensor data (Rodseth & Burmeister, 2015; DNV, 2021).

For example, GPS finds a ship's location, sonar finds underwater dangers, AIS finds other ships nearby, and smart sensors keep an eye on engine performance and environmental conditions. When put together, these data streams let AI-driven systems figure out the safest and most efficient ways to get around (Schill & Browning, 2015; Haugen & Kristiansen, 2022).

Benefits to Maritime Operations

1. **Enhanced Navigational Accuracy** - Multi-source data integration ensures redundancy and reliability.
2. **Improved Safety** - Early detection of hazards both above and below the waterline.
3. **Operational Efficiency** - Optimization of speed, routing, and maintenance planning.
4. **Regulatory Compliance** - Facilitates adherence to safety, environmental, and reporting requirements (IMO, 2021; Durlik et al., 2024).

These integrated technologies are the foundation of AI-enabled maritime intelligence. Even the best AI algorithms can't work well without accurate, timely, and reliable data from GPS, sonar, AIS, and smart sensors. As marine innovation progresses, the seamless amalgamation of these sensing technologies with AI will be essential for attaining fully autonomous, secure, and sustainable maritime operations (Grand View Research, 2024; SaveOurBlueOcean, n.d.).

There are four parts that work together to make AI-driven maritime intelligence possible: machine learning algorithms, real-time analytics with predictive modeling, autonomous control systems, and integrated sensor technologies. These parts work together to make operations safer, more efficient, and more environmentally friendly, which will eventually lead to the development of next-generation intelligent and autonomous maritime operations (Durlik et al., 2024; Zhu et al., 2025).

3. APPLICATIONS FOR AI IN MARITIME ACTIVITIES

Artificial intelligence has gone from being tested to being a valuable tool that people in the maritime industry may use every day. AI applications make everything from planning trips to preserving the environment safer, more efficient, more compliant, and more sustainable. These changes will help with both old concerns, like safety on the sea and port congestion, and new ones, including climate change and cybersecurity.

3.1. Navigation and Route Optimization

Navigation is a key part of marine operations, and recent improvements in artificial intelligence (AI) have made it possible to plot routes more accurately and quickly. AI-powered systems get around these problems by using real-time data from several sources, such as weather forecasts, ocean currents, tidal patterns, vessel traffic, environmental laws, and reports of port congestion (DNV, 2021; IMO, 2021). With these inputs, the algorithm can figure out the safest, most fuel-efficient, and most legal routes. AI-based navigation, on the other hand, keeps updating routes during journeys to account for changes in the environment and operations.

AI-based navigation lets you make changes on the flight while you're on your trip. For example, a system may suggest changing speed before a storm or changing routes to take advantage of good currents while keeping the timetable. This flexibility lowers risks, expenses of doing business, and damages the environment (Rodseth & Burmeister, 2015; Durlik et al., 2024).

Benefits of AI-Based Route Optimization

1. **Fuel Savings and Cost Reduction -** Optimized routing can reduce fuel consumption by 3–10%.
2. **Emission Reduction -** Supports compliance with IMO decarbonization targets by lowering greenhouse gas emissions.
3. **Enhanced Safety -** Early identification of hazards and traffic conflicts reduces accident risk.
4. **Improved Schedule Reliability -** More accurate ETA predictions improve port coordination and supply chain efficiency.

Big shipping businesses and service providers are starting to use routing systems that use AI. For instance, Maersk uses weather routing algorithms that use satellite data, ship performance, and oceanographic models to make transoceanic trips as smooth as possible. StormGeo also offers routing services based on machine learning to thousands of ships around the world, which makes them safer to operate and use less fuel (Grand View Research, 2024).

3.2. Cargo Management and Port Operations

To keep global supply chains going, ports and cargo management need to be done well. Trade networks can be quite expensive to interrupt if scheduling, handling, or logistics are not done on time. Artificial intelligence (AI) makes these procedures better by automating scheduling, making tracking more accurate, and optimizing logistics at both the shipboard and port levels (Rodseth & Burmeister, 2015; Haugen & Kristiansen, 2022).

AI in Cargo Management Onboard

AI-powered systems keep an eye on cargo conditions, make the most of storage space, and make sure that loads are distributed evenly for safety and stability.

1. **Container Tracking -** AI integrates RFID, GPS, and sensor data to provide real-time visibility of container location and condition. This includes monitoring temperature-sensitive goods, detecting vibrations, and identifying potential damage during transit.
2. **Load Planning and Optimization -** Algorithms determine the most efficient cargo arrangement to maximize capacity while maintaining vessel stability and safety compliance.
3. **Risk Mitigation -** Predictive analytics identify containers at risk of delay or damage, enabling proactive interventions before disruptions occur (DNV, 2021).

AI in Port Operations

Ports are highly complex environments requiring efficient coordination of resources. AI applications include:

1. **Automated Berth Scheduling -** Systems analyze vessel arrival times, cargo types, and port resources to allocate berths effectively, minimizing congestion and delays.
2. **Optimized Equipment Allocation** - Machine learning models forecast peak workloads and schedule cranes, trucks, and personnel to reduce downtime.
3. **Intermodal Transfer Management -** AI facilitates the seamless movement of cargo between ships, trucks, and trains, enhancing overall logistical efficiency (Durlik et al., 2024).

Integration with Predictive Analytics

Predictive models help with operational planning by taking into consideration bad weather, a lack of workers, or traffic jams at transportation hubs. This lets port owners change the schedules of their workers, send goods to other terminals, and tell shipping lines and clients about new arrival dates (Grand View Research, 2024).

Benefits of AI in Cargo and Port Management

1. **Reduced Delays -** Automated scheduling accelerates vessel turnaround.
2. **Increased Throughput -** Optimized resource allocation enables higher cargo volumes within shorter timeframes.
3. **Cost Efficiency -** Improved utilization of assets lowers operational costs.
4. **Enhanced Transparency -** Real-time tracking fosters greater trust among shipping lines, ports, and customers.

AI-powered scheduling and container management systems at the Port of Singapore handle millions of data points every day to help with effective berth allocation and crane operations. This makes it possible to move more than 37 million TEUs each year with little traffic. APM Terminals also uses AI to lay out yards and stack containers automatically, which speeds up the time it takes for ships and trucks to turn around (Haugen & Kristiansen, 2022; Grand View Research, 2024).

3.3. Maritime Threat Detection and Security Enforcement

For global trade and national defense, it is very important to keep the seas safe, especially since there are always threats like piracy, smuggling, human trafficking, and illegal fishing. Manual patrols, radar monitoring, and visual inspections are some of the traditional methods used to find dangers early enough to stop them from getting worse. Artificial intelligence (AI), along with computer vision, sensor fusion, and predictive modeling, makes surveillance better and helps security enforcement be more proactive (Rødseth & Burmeister, 2015; DNV, 2021).

AI-Driven Surveillance Systems

AI and computer vision are used by modern surveillance systems to analyze data from cameras, drones, radars, and satellites.

1. **Piracy Detection -** AI algorithms analyze ship behaviors and imagery to identify small, fast-moving vessels approaching commercial ships in high-risk zones such as the Gulf of Aden.
2. **Smuggling Prevention -** Systems integrate AIS data with video and radar feeds to detect vessels operating without proper identification or deviating from approved routes.
3. **Illegal Fishing Detection -** Satellite-based AI platforms, such as Global Fishing Watch, monitor fishing vessels for unusual patterns, including extended activity in protected areas (Grand View Research, 2024).

Benefits of AI in Maritime Security

1. **Faster Detection and Response** - Reduces the time between threat identification and intervention.
2. **Expanded Surveillance Coverage** - Satellite and UAV integration enables monitoring of large and remote oceanic regions.
3. **Reduced False Alarms** - Intelligent data filtering improves accuracy in distinguishing genuine threats from benign behaviors.
4. **Support for Law Enforcement and Compliance** - Facilitates enforcement of maritime boundaries, fishing regulations, and anti-smuggling laws.

The European Maritime Safety Agency (EMSA) uses AI-powered satellite monitoring to find illegal fishing, pollution, and strange conduct by ships in EU waterways. Private sector platforms like Windward AI also utilize predictive analytics to find "dark ships," which are ships that turn off their AIS transponders to avoid being found. This makes coast guard and naval patrols more effective (Durlik et al., 2024).

3.4. Environmental Monitoring and Sustainability Initiatives

The maritime industry has two roles: it helps the global economy flourish and also puts a lot of stress on the environment, such as by releasing greenhouse gases, polluting the ocean, and destroying habitats. Shipping alone is responsible for more than 3% of global greenhouse gas emissions. Maritime activities also cause oil spills, plastic waste, and the loss of biodiversity (DNV, 2021; Haugen & Kristiansen, 2022). Artificial intelligence (AI) gives us tremendous tools to keep an eye on these environmental effects and support sustainable practices that follow international rules, including the International Maritime Organization's (IMO) plan to cut carbon emissions.

AI in Oil Spill Detection

Oil spills pose long-term risks to ecosystems, requiring rapid identification and response.

1. Satellite and Aerial Image Analysis – Computer vision models detect oil slicks in high-resolution imagery, even under cloudy or low-light conditions.
2. Automated Alerts – AI systems provide real-time notifications to relevant authorities, expediting containment and cleanup efforts (Grand View Research, 2024).

Tracking Marine Pollution

AI technologies extend beyond oil spill detection to monitor diverse forms of pollution, including plastics, chemicals, and wastewater.

1. **Plastic Waste Recognition** - Drone and satellite imagery analyzed by AI can identify and map floating debris.
2. **Pollution Source Tracing** - Algorithms integrate vessel activity data with pollution detection events to identify potential sources of illegal discharges.
3. **Predictive Modeling** - Analytics estimate the dispersion of pollutants, enabling prioritization of cleanup operations (Durlik et al., 2024).

Monitoring Coral Reef Health

Coral reefs, critical for biodiversity, face increasing risks from climate change and human activity.

1. **Underwater Imaging and AI Analysis** - AI models process imagery from remotely operated vehicles (ROVs) and autonomous underwater drones to assess reef conditions and identify bleaching events.
2. **Automated Habitat Mapping** - Deep learning systems generate detailed maps showing changes in coral cover, species richness, and structural complexity over time (SaveOurBlueOcean, n.d.).

Integration with Sustainability Initiatives

AI-driven monitoring supports broader sustainability goals in the maritime industry:

1. **Emission Tracking** - Sensors on ships measure fuel consumption and exhaust output, with AI ensuring compliance with emission control standards.
2. **Sustainable Route Planning** - Navigation algorithms select eco-friendly routes that reduce fuel use and avoid ecologically sensitive areas.
3. **Fisheries Management** - AI systems support enforcement of catch limits and help deter illegal, unreported, and unregulated (IUU) fishing (IMO, 2021).

Benefits of AI in Environmental Sustainability

1. **Early Detection of Threats** - Enables rapid mitigation of ecological hazards.

2. **Data-Driven Conservation** - Provides evidence-based insights for policymakers and regulators.
3. **Regulatory Compliance -** Assists stakeholders in meeting international environmental standards.

AI applications in environmental monitoring show how technology may change the maritime industry into a force for sustainability. AI helps people follow international rules by allowing them to find oil spills, track pollution, track coral reefs, and use eco-friendly methods of doing business. This also helps people reach their long-term conservation goals. As these technologies improve, the maritime industry provides a potential shift from being a source of environmental strain to becoming an active partner in protecting the world's oceans (Haugen & Kristiansen, 2022; DNV, 2021).

4. ADVANTAGES AND STRATEGIC BENEFITS

Nowadays, the modern maritime industry makes the ability to provide timely and accurate decisions increasingly vital for operational efficiency, safety, and competitiveness. Traditional decision-making in the maritime industry relies on experience, intuition, and manual reporting, which can be in dynamic contexts characterized by unpredictable weather, complex logistics, and high vessel traffic. Maritime stakeholders may make choices faster by leveraging modern data analytics, AI, and real-time monitoring. This gives them both operational and strategic benefits.

4.1. Enhancing Decision-Making Speed and Accuracy

Making decisions about maritime operations is both important and very hard. Weather, port congestion, vessel traffic, and cargo logistics are always changing, which makes things difficult for captains, fleet operators, and port authorities. So, making judgments quickly and correctly is very important for keeping the sector safe, efficient, and competitive.

Importance of Speed in Decision-Making

Delays in decision-making can lead to a chain reaction that affects safety, scheduling, and operational costs. However, with the help of modern technology, the processing of large operational datasets now allows maritime stakeholders to:

- Accuracy Through Data-Driven Insights

- Accuracy is equally critical, as informed decisions help prevent errors, reduce accidents, and optimize resources.
- Analyzing historical and real-time datasets to predict risks and delays.
- Identifying optimal routing and operational practices.
- Enabling predictive maintenance to minimize unexpected equipment failures.

For instance, the use of AI-powered predictive analytics can alert captains to potential hazards during a voyage and assist port authorities in assigning docks more efficiently based on real-time vessel traffic.

Strategic Benefits

The maritime industry benefits a lot from making decisions more quickly and accurately. For captains, it makes navigation safer, helps them plan the best routes, and makes them better at responding to emergencies. For fleet operators, it helps them manage their fleets more efficiently, decreases their operating costs, and speeds up the delivery of cargo. It helps port management plan better, move more cargo through the port, and lessen traffic.

Addressing the challenges that hinder these advancements is essential to maximizing the potential of AI and data-driven decision-making. By overcoming such barriers, maritime stakeholders can better adapt to changing conditions and fully utilize evidence-based systems that enhance safety, efficiency, and overall performance.

4.2. Improving Safety and Reducing Human Error

Safety relics are the leading precedence in maritime operations, with human error depending on one of the leading causes of mishaps at sea (International Maritime Organization [IMO], 2021; European Maritime Safety Agency (2022). The maritime environment presents unique risks, including severe weather, complex navigation, heavy machinery, and high-value cargo, which amplify the consequences of human mistakes (Bal Beşikçi et al., 2019). By ritual, vessel safety has relied heavily on crew expertise and vigilance. However, miscalculations in navigation, communication disappointments, and lapses in decision-making continue to result in significant incidents. Recent developments in automation, artificial intelligence (AI), and digital decision-support systems are transforming safety management by reducing human error and strengthening operational reliability (Mallam, Nazir, & Sharma, 2020).

The Role of Technology in Reducing Human Error

Emerging technologies are capable of supporting or partially automating critical functions, thereby reducing dependence on human judgment alone. Key applications include:

1. **Automated Navigation Systems** - Deliver real-time route optimization, collision alerts, and course corrections to reduce navigational risks (Zhou et al., 2020).
2. **Predictive Maintenance Tools** - Monitor engine and equipment performance, providing early warnings of potential failures (Lu et al., 2021).
3. **AI-Powered Decision Support** - Analyze extensive datasets to assist captains and port operators in making safer, evidence-based decisions (Wang & He, 2021).
4. **Sensor Networks and IoT Devices** - Enable continuous monitoring of vessel conditions, cargo security, and environmental parameters, issuing timely hazard alerts (Yang, Xu, & Li, 2019).

Strategic Advantages

Stakeholders in the maritime industry can gain a competitive edge and make things safer by reducing human error. Reduced incidents mean better logistics, less downtime, and more efficient use of the fleet, all of which contribute to operational efficiency (Zhou et al., 2020). Preventing accidents and equipment failures saves money since it cuts down on costs for repairs, insurance premiums, and fines from the government (Yang, Xu, & Li, 2019). A good safety record builds trust and reputation, which makes you more credible with regulators, clients, and worldwide partners (Mallam, Nazir, & Sharma, 2020).

Challenges to Implementation

Even with these benefits, there are still a number of problems that make it hard for the marine industry to fully implement AI. Training and adaptation are necessary since crew members need to learn how to use and trust automated technologies (Mallam, Nazir, & Sharma, 2020). Reliability of the system is another issue since as we rely more on digital systems, we need strong backup systems to keep things running smoothly (Lu et al., 2021). Cybersecurity hazards are also very dangerous because the Internet of Things (IoT) and other linked technologies make us more vulnerable to cyberattacks that could put our safety at risk (Yang, Xu, & Li, 2019). A strategic approach to adopting these technologies not only safeguards human life but also strengthens resilience and reliability within the maritime industry (IMO, 2021; European Maritime Safety Agency, 2022).

4.3. Increasing Operational Efficiency and Cost-Effectiveness

For the maritime business to be competitive, it is important to be efficient and cost-effective. Efficient operations help avoid delays, make the best use of resources, and cut down on waste, while cost-effectiveness keeps the business financially stable without compromising safety or service quality (Bal Beşikçi et al., 2019; Mallam, Nazir, & Sharma, 2020). The use of new technologies like AI, automation, and real-time analytics is changing how things are done at sea by making workflows more efficient, cutting down on mistakes made by people, and helping people make decisions based on facts (IMO, 2021).

To be effective in the marine industry, vessel operations, fleet management, cargo handling, and port activities all need to be as efficient as possible. Predictive analytics are very important for making route planning better since they find the safest and most fuel-efficient paths, which cuts down on travel time and expenses (Zhou, Xu, & Li, 2020). Real-time monitoring solutions make fleet and resource management even better by letting operators assign vessels, crew, and cargo more efficiently and cut down on downtime. Automation is also changing how ports work by using improved scheduling and dock allocation systems that cut down on traffic and make things run more smoothly (European Maritime Safety Agency, 2022). Predictive maintenance solutions also make things more reliable by finding possible equipment faults before they happen, which saves expensive breakdowns and makes sure that performance stays the same (Lu, Liu, & Yang, 2021). All of these improvements show how important digital transformation and AI-driven solutions are for making the maritime supply chain more efficient, cutting costs, and making it more sustainable.

Improving Cost-Effectiveness

Technology-driven strategies help reduce costs while maintaining service quality. Some of the key benefits include:

1. **Fuel Efficiency -** Engine monitoring and optimized routing reduce fuel consumption, the largest cost in maritime transport (Yang, Xu, & Li, 2019).
2. **Error Minimization -** Automation prevents costly mistakes in navigation and cargo handling.
3. **Reduced Maintenance Costs -** Predictive tools extend equipment lifespan and minimize unplanned repairs.

4. **Resource Optimization -** AI analytics improve workforce deployment and cargo distribution, reducing operational expenses.

Strategic Advantages

Enhanced efficiency and cost-effectiveness yield broader benefits:

1. **Competitive Advantage** - Faster, more reliable services strengthen market reputation.
2. **Sustainability -** Lower fuel use and optimized operations contribute to compliance with environmental regulations (International Maritime Organization, 2021).
3. **Scalability -** Efficient processes support expansion without proportional cost increases.
4. **Profitability -** Reduced expenses and optimized throughput lead to stronger financial stability.

Challenges and Considerations

Implementation requires addressing several challenges:

1. **Data Integration -** Optimization depends on accurate, real-time information across vessels, ports, and logistics chains.
2. **Technology Investment -** High upfront costs may deter adoption, though long-term savings are significant.
3. **Training and Change Management -** Personnel must be equipped to operate innovative systems effectively.
4. **Cybersecurity -** Networked systems must be protected against cyber threats (Mallam, Nazir, & Sharma, 2020).

In summary, increasing operational efficiency and cost-effectiveness is essential for maritime affordability. Through AI, predictive analytics, and computerization, stakeholders can modernize developments, optimize resources, and achieve maintainable financial performance while meeting regulatory and safety obligations (International Maritime Organization, 2021; European Maritime Safety Agency, 2022).

4.4. Supporting Global Maritime Security

To keep trade routes, ships, and national interests, a global nautical refuge is very important. Piracy, smuggling, illicit fishing, and geopolitical disputes are still problems in the marine sector (International Maritime Organization, 2021; European Maritime Safety Agency, 2022). Keeping things safe not only protects the crew and cargo, but it also helps the economy and international stability. To solve these problems, we need more advanced technology, such as artificial intelligence (AI), satellite observing, and integrated surveillance systems.

The Role of Technology in Enhancing Security

Technological revolution enables maritime specialists to monitor large areas efficiently, detect threats early, and respond rapidly. Key applications include:

1. **Surveillance and Monitoring Systems -** Satellites, drones, and radar networks continuously track vessel movements, identifying anomalies.
2. **AI-Powered Threat Detection -** Machine learning algorithms analyze behavioral patterns to flag piracy, smuggling, or illegal fishing activities (Rodseth & Burmeister, 2015).
3. **Cybersecurity Measures -** Protecting routing, port operations, and communication networks eases digital vulnerabilities (Mallam, Nazir, & Sharma, 2020).
4. **Integrated Security Platforms -** Fusing data from AIS, radar, and satellite sources supports organized responses between navies, coast guards, and port authorities worldwide (Windward, 2021).

These tools enhance situational awareness, allowing authorities to prevent and respond to threats more effectively.

Strategic Importance of Maritime Security

Robust maritime security yields significant benefits:

1. **Protection of Global Trade** - Certifies the continual measure of goods, vital for commercial stability.
2. **Safety of Crew and Cargo -** Reduces risks from piracy and illicit maritime activity.
3. **National and Regional Security -** Safeguards territorial waters and strategic chokepoints.
4. **International Cooperation -** Promotes joint initiatives through intelligence sharing and multilateral frameworks (International Maritime Organization, 2021).

Challenges to Maritime Security

Despite progress, several challenges persist:

1. **Resource Limitations** - Traversing vast water requires substantial investment.
2. **Evolving Threats** - Criminal groups continuously adapt, demanding flexible and intelligent systems.
3. **International Coordination** - Jurisdictional differences can hinder joint responses.
4. **Cyber Vulnerabilities** - Growing reliance on digital systems introduces risks of cyberattacks.

To make shipping secure and effective worldwide, it is vital to support global marine security. By uniting AI-driven monitoring, predictive threat detection, and cybersecurity measures, stakeholders can lower risks, protect crews and cargo, and make global business networks more resilient (EMSA, 2022; Windward, 2021). Nautical security protects not only economic interests but also encourages cooperation between countries and stability in the interconnected marine environment.

5. CHALLENGES AND LIMITATIONS

The maritime industries are currently restructuring with the use of Artificial Intelligence (AI) in the maritime sector, using predictive analytics, some automation, and enhanced situational awareness. The integration of AI can bring safety, efficiency, and security across maritime operations. However, these advantages are safely accompanied by several challenges and transitions that must be addressed to ensure responsible and sustainable implementation. The embracing of AI to become effective requires confronting its several keys like, technical, financial, regulatory, cybersecurity, and ethical limitations (Vallero, Renga, Meo, & Ajmone Marsan, 2020; Mallam, Nazir, & Sharma, 2020).

AI models are based on heavily quality of training data. According to Vallero, et.al., (2020) unknown events may result in inaccurate predictions that can have big impact on training data, which is one of the primary obstacles lie in the technical limitations of legacy infrastructures. To lessen the risks in forecasting data, a regular testing system must be practiced on ships and make sure to secure backup measures with a strong cybersecurity framework that should be required. Reliability can also be conceded by hardware failures, it can also affect sensor malfunctions, or maybe software bugs which can be hassled to identify errors, while the increased connectivity of maritime systems exposes them to cybersecurity risks that can disrupt operations or manipulate decision-making outputs (Mallam, Nazir, & Sharma, 2020).

This integration of AI into vessel operations presents both technical challenges and financial investment for the owners. AI platforms used advanced sensors and observing systems, it requires ongoing recalibration testing, ship improvements, and technical support to maintain its performance. However, the disadvantage of using AI during at sea is it has limited access to reliable internet connections and unstable satellite communication, particularly in remote vessel region which it restricts common distribution (European Maritime Safety Agency, 2022), in terms of financial and industry-related challenges it makes difficult for developing economies to adopt such new systems as this.

Moreover, accountability in vessel operations have some issues which remain unresolved, it determines the responsibility and liability of AI driven operational navigational decisions in a complex challenge. Regulatory and policy gaps are also one of the major obstacles in adopting AI. The lack of universally recognized ethical standard for AI in maritime sector which creates fragmented regulatory landscape in maritime systems, as countries have varying rules in data governance, vessel automation, and cybersecurity (International Maritime Organization, 2021).

6. HUMAN FACTORS, ECONOMIC ROI, AND GOVERNANCE

Adopting effective upskilling programs and promoting human-machine collaboration are important for a workforce transition plan using AI in naval operations. This shift involves technical implementations that also impact human factors. Maritime staff training program needs change into the adaptability of automation and the psychological impact of reduced human control remain an important or yet often overlooked concern.

Governance and interoperability add another layer of complexity to current challenges. Inconsistent regulatory standards, data-sharing hesitancy, and uneven cybersecurity practices hinder global synchronization. Future frameworks should clarify liability in the maritime system, the implementation of ethical responsibility, and the collaboration of international organizations to ensure that AI advances benefit the entire maritime sector.

7. FUTURE TRENDS: BLOCKCHAIN, QUANTUM COMPUTING, AND BEYOND

Today, the rapid pace of the advancement of technology has greatly influenced the use of AI in the shipping industry. Blockchain for example, holds and secures logistics chains that increases transparency in shipping paperwork tasks and prevents possible fraud. Although it is still in its early stages, the quantum computing holds promise for tackling complex optimization in route planning and coordination in the supply chain at extraordinary speeds. Integrating sensors with Hybrid AI and IoT frameworks is expected to enhance predictive maintenance capabilities, while edge computing will enable quick decision-making at sea without the stipulation of constant connectivity. AI systems in the maritime sector are becoming autonomous but also more transparent, secure, and globally integrated.

CONCLUSION

The chapter is typically well-ordered, initial with an indication of AI-driven marine intelligence and then going over its benefits, uses, and problems. For example, predictive maintenance and route optimization are discussed in more than one part, which makes the chapter feel like it goes in circles in its place of straight lines. Today, many sections tend to overlap on focusing on cost-effectiveness and decision-making which makes it difficult for the ideas to flow smoothly and cohesively.

By managing a complex task, the AI technologies provide help in industries worldwide to operate smoothly in minimizing risks and ensuring compliance with safety and security regulations.

The use of AI in maritime operations won't make it a problem. There are rules and guidelines that are established to aim on how to improve cybersecurity and data privacy in addressing the issues related to integration and data quality and managing challenges arising from registered company systems. These thoughts are important to ensure the safe and responsible use of AI int the maritime industry.

The emerging system of AI driven in maritime intelligence is designed to enhance and improve the management of shipping and port operations. By this, it combines human specialty with machine use intelligence to make maritime activities safer and more efficient and better connected on a global standard.

REFERENCES

Almutairi, A., & Khan, R. (2023). Image-based classical features and machine learning analysis of skin cancer instances. *Applied Sciences (Basel, Switzerland), 13*(13), 7712. DOI: 10.3390/app13137712

Bal Beşikçi, E., Kececi, T., Arslan, O., & Turan, O. (2019). An artificial neural network- based decision support system for maritime safety management. *Safety Science, 120*, 490–500. DOI: 10.1016/j.ssci.2019.07.031

DNV. (2021). *Maritime autonomous surface ships: Regulatory framework and challenges* [Technical and regulatory news]. DNV. Retrieved from https://www.scribd.com/document/516077075

Durlik, I., Miller, T., Kostecka, E., & Tuński, T. (2024). Artificial intelligence in maritime transportation: A comprehensive review of safety and risk management applications. *Applied Sciences (Basel, Switzerland), 14*(18), 8420. DOI: 10.3390/app14188420

Grand View Research. (2024). *Maritime artificial intelligence market: Industry report, 2030*. Grand View Research. https://www.grandviewresearch.com/industry analysis/maritime-artificial-intelligence-market-report

Hale, J., Long, S., Corns, S., & Shoberg, T. (2019). Flood management deep learning model inputs: A review of necessary data and predictive tools. *Proceedings of the International Annual Conference of the American Society for Engineering Management (ASEM)*, 1–9. DOI: 10.5465/AMBPP.2019.15761

Haugen, S., & Kristiansen, S. (2022). *Maritime transportation: Safety management and risk analysis* (2nd ed.). Routledge., DOI: 10.4324/9781003055464

International Maritime Organization. (2021). World maritime theme: Seafarers at the core of shipping's future. IMO. https://www.imo.org

International Maritime Organization. (2025, March 18). *IMO to develop a global strategy for maritime digitalization*. https://www.imo.org/en/MediaCentre/PressBriefings/pages/IMO-global-strategy maritime-digitalization.aspx

Lu, Y., Liu, C., Wang, K. I.-K., Huang, H., Xu, X., & Zhang, W. (2021). Industrial Internet of Things (IIoT): A smart manufacturing perspective. *IEEE Transactions on Industrial Informatics, 17*(3), 2390–2402. DOI: 10.1109/TII.2019.2942495

Mallam, S. C., Nazir, S., & Sharma, A. (2020). The human element in future maritime operations—Perceived impact of autonomous shipping. *Ergonomics, 63*(3), 334–345. DOI: 10.1080/00140139.2019.1659998

Mohammed, M. A., & Ahmed, A. (2025). Artificial intelligence applications in cancer diagnostic devices: A focused review. *Medical Oncology Journal*. Advance online publication. DOI: 10.61706/MECOJ160160

Pacific Data Integrators. (n.d.). *PDI customer care: Revolutionize financial services with generative AI*. Pacific Data Integrators. https://www.pacificdataintegrators.com/industry solution/financial-services-with- generative-ai

Rodseth, O. J., & Burmeister, H.-C. (2015). *Risk assessment for autonomous ships. Transactions on Maritime Science, 4*(1), 31–40. Retrieved from https://www.transnav.eu/Article_Risk_Assessment_for_an_Unmanned_R%C3% B8dseth %2C35%2C593.html

Sameera Sulakshana, B. K. (2024). *New technologies to enhance GMDSS and reduce workload on deck officers* [PDF]. CORE. https://core.ac.uk/download/638723085.pdf

SaveOurBlueOcean. (n.d.). *Encouraging the adoption of sustainable shipping practices to minimize marine pollution*. Retrieved August 2025, from https://www.saveourblueocean.shop/post/encouraging-the-adoption-of-sustainable shipping-practices-to-minimize-marine-pollution

Schill, N., & Browning, M. C. K. (2015). The impact of satellite AIS to the environmental challenges of modern shipping. In *Maritime Informatics* (pp. 1–14). Institution of Engineering and Technology. DOI: 10.1049/PBTR001E_ch13

Vallero, G., Renga, D., Meo, M., & Ajmone Marsan, M. (2020). Processing ANN traffic predictions for RAN energy efficiency. In *Proceedings of the 23rd International ACM Conference on Modeling, Analysis and Simulation of Wireless and Mobile Systems* (MASWiM '20). ACM. DOI: 10.1145/3416010.3423222

Wang, T., & He, J. (2021). Artificial intelligence applications in maritime transportation: A review and future research directions. *Ocean Engineering, 237*, 109680. DOI: 10.1016/j.oceaneng.2021.109680

Winward, M. (2021). Intelligence capacity and mass violence: Evidence from Indonesia. *Comparative Political Studies, 54*(3-4), 553–584.

Xia, W., Xing, K., & Guo, Z. (2022). Target tracking method of shipborne navigation radar. In J. C. Hung, N. Y. Yen, & J. W. Chang (Eds.), *Frontier Computing* (Lecture Notes in Electrical Engineering, Vol. 827) (pp. ...). Springer Singapore. DOI: 10.1007/978-981-16-8052-6_262

Yang, C., Xu, Q., & Li, Z. (2019). Ship safety monitoring system based on the Internet of Things. *Journal of Coastal Research*, *93*(sp1), 698–702. DOI: 10.2112/SI93-094.1

Zhou, X., Zhang, L., & Qiu, Z. (2020). Intelligent navigation system for autonomous ships based on data fusion and deep learning. *Applied Ocean Research*, *95*, 102020. DOI: 10.1016/j.apor.2019.102020

Zhu, H., Wang, J., & Gao, R. (2025). EEG-based inverse reinforcement learning for safety- oriented global path planning in dynamic environments. *Applied Sciences (Basel, Switzerland)*, *15*(11), 6163. DOI: 10.3390/app15116163

Chapter 5
Regulatory, Ethical, and Economic Considerations

S. Ida Evangeline
https://orcid.org/0000-0003-2997-7897
Government College of Engineering, Tirunelveli, India

Abstract

As autonomous vessels and maritime drones transition from experimental technologies to operational realities, they are redefining the structural, regulatory, and ethical foundations of the global maritime industry. While these innovations promise enhanced efficiency, safety, and environmental performance, they also present a range of complex challenges that extend beyond technical feasibility. This chapter provides a holistic, interdisciplinary analysis of the regulatory, ethical, and economic issues of maritime automation. It examines how international frameworks, especially the ones shaped by the International Maritime Organization, are struggling to come to terms with the implications of autonomy. It also identifies some of the most critical gaps in legal accountability, certification, and cybersecurity. Ethically, the chapter questions the transfer of navigational responsibility to machines, the possibility of replacing maritime labor, and dangers of opaque or prejudiced artificial intelligence systems. Economically, it measures the transformative potential and the disruptive effects of automation, especially focusing on capital costs, labor market changes, and new business models. By means of systems-level analysis, the chapter highlights the interdependence of these domains and positions for anticipatory, inclusive governance frameworks. It ends up with concrete recommendations for policymakers, technologists, and industry leaders on how maritime innovation can be technologically sound, ethically grounded, and economically sustainable.

DOI: 10.4018/979-8-3373-3546-9.ch005

1. INTRODUCTION

The maritime industry is entering a period of profound transformation, catalyzed by the rapid advancement and integration of emerging technologies such as autonomous surface vessels, unmanned underwater and aerial drones, machine learning systems, and advanced sensor networks. Such innovations are transforming how maritime activities are envisaged, planned, and carried out – from transoceanic shipping and port logistics to maritime surveillance, environmental monitoring, and emergency response. As the sector moves toward digitization and automation, it will gain operational efficiency, reduce human error, increase environmental compliance, and create new commercial opportunities (Farah et al. 2024).

However, the pace of technological innovation carries with it a slew of intricately linked difficulties that go far beyond technical considerations. Regulatory frameworks must adapt and regulate new types of vessel control and decision-making. Ethical concerns arise as autonomous systems replace human judgment in high-risk maritime environments (Pareek et al. 2025). Economic models must adapt to reflect the shifting cost structures, market dynamics, and labor effects of automation. These characteristics are not ancillary; rather, they are critical to the responsible and sustainable deployment of developing maritime technologies (Animah, 2024).

This chapter will critically explore these three interconnected pillars: regulation, ethics, and economics. Each is a perspective through which the transformational potential of maritime automation should be assessed and developed. Regulatory analysis is concerned with the dynamic legal and institutional responses to maritime autonomy, such as those provided by the International Maritime Organization and other national agencies. It calls into question whether existing conventions and norms are ready to adopt technologies that fundamentally challenge beliefs about vessel control, personnel requirements, and legal accountability.

The ethical dimension addresses issues of accountability, openness, and fairness. Autonomous maritime systems must be programmed to do actions that are not only lawful but also morally justifiable, particularly in situations involving human life, environmental hazards, or socioeconomic disturbance. Ethical concerns have migrated to the design processes and decision-making logics incorporated into artificial intelligence, raising problems about which values are important and which dangers are externalized.

Economically, the adoption of autonomous maritime technologies has the potential to rebalance world shipping and trade. The predicted reduction in operating expenses, logistical optimization, and the development of new business models may deliver significant benefits. However, these benefits are outweighed by high initial expenses, ambiguities in economic returns, and the social costs of worker displacement and structural changes in maritime jobs. Furthermore, international gaps in finance and technological knowledge might exacerbate the divide between first-tier maritime countries and coastal emerging economies.

By combining concepts from legal research, applied ethics, and economic systems theory, this chapter seeks to provide a comprehensive and multidisciplinary framework for understanding how maritime innovation might be ethically controlled. The end product is not simply a description of the change that is taking place, but also the tools and concepts that stakeholders may use to shape it. This chapter contributes to the overarching goals of this collection by adding to the collaborative effort of mapping a future in which technology advances and societal values are balanced in the global marine domain.

2. REGULATORY LANDSCAPE

The incorporation of autonomous technologies into the maritime industry calls into question the fundamental tenets of current regulatory frameworks. Traditional maritime governance is built on the notion of human agency, with shipmasters and personnel accountable for navigation, safety, communication, and adherence to international conventions. The introduction of autonomous surface vessels, unmanned aerial and undersea systems, and algorithm-driven operating procedures challenges this paradigm, necessitating a legislative overhaul that is both technologically informed and future-oriented.

At the worldwide level, the International Maritime Organization (IMO) has acknowledged the importance of resolving this regulatory shift. Its ongoing efforts, including the Maritime Autonomous Surface Ships (MASS) regulatory scoping exercise, are to review and evaluate existing international conventions such as the International Convention for the Safety of Life at Sea (SOLAS), the International Regulations for Preventing Collisions at Sea (COLREGs), and the Standards of Training, Certification, and Watchkeeping for These basic devices were designed in an atmosphere in which the presence of humans on board the ship was assumed. In this way, they carry implicit assumptions - such as the availability of a watchkeeping officer, the use of visual cues to navigate, or a human's ability to intervene immediately - that may not be applicable to autonomous operations (Ahmed et al. 2024).

The IMO's scoping work is an important first step, but it remains primarily experimental. The shift from evaluation to regulation would necessitate unprecedented cooperation among member states, many of whom have opposing strategic objectives, technological capabilities, and marine traditions. In the meanwhile, national and regional entities are beginning to fill the void with experimental and provisional policies (Reddy et al. 2025). The European Union has supported regulatory innovation through research consortiums such as MUNIN (Maritime

Unmanned Navigation using Intelligence in Networks) and AUTOSHIP, which test not just new technology but also associated governance frameworks. Countries like Norway, Japan, and Singapore have developed regulated testing areas and interim legal structures to facilitate early implementation, but others are behind or employing outdated legislation inadequate for digital marine operations (Wang et al. 2024).

Among the most pressing issues for regulatory development are vessel registration and flagging. The current system requires that a ship be registered under a national flag and subject to that nation's regulatory scrutiny. Autonomous ships raise challenging concerns concerning what constitutes a "flag state" when crew and command organizations are decentralized or nonexistent. The similar ambiguity arises with liability and insurance. If an autonomous ship causes harm, who is responsible: the software developer, the ship's owner, the operator, or the algorithm? The lack of open models of accountability impedes both legal action and the development of marine insurance solutions that are fit for autonomy (Hebbar, 2024).

Certification and categorization criteria are another key issue. Classification bodies have set rules for inspecting and certifying ships as seaworthy. These criteria for autonomous systems need to be revised to include not only mechanical dependability, but also software integrity, fail-safe mechanisms, cyber resilience, and decision-making logic. Although classification societies like Lloyd's Register and DNV are beginning to investigate these dimensions, a single global strategy remains out of reach (Voroniuk, 2024).

Cybersecurity has also emerged as a regulatory frontier. Autonomous vessels rely significantly on networked technologies for navigation, propulsion, communication, and remote monitoring. These systems are vulnerable to cyberattacks, such as GPS spoofing, malware, and fleet control network attacks. Regulatory authorities must thus adopt stringent cybersecurity criteria like as encryption methods, access control systems, and real-time threat detection capabilities. These activities must be proactive rather than reactive, as the maritime realm becomes an increasingly appealing target for cybercriminals and geopolitical adversaries (Potamos et al. 2024).

Collision avoidance is arguably the most technically and ethically demanding domain. COLREGs were designed to be read by trained human navigators using their judgment in real-world settings. The technical complexity of translating these rules into machine-readable algorithms is not the only challenge, but it also involves normative interpretation. How should a machine balance safety of self against safety of others and environment? Regulatory guidance thus has to go beyond formal rules to include acceptable decision-making limits for autonomous systems, especially in the mixed environments where human-piloted and unmanned vessels operate together (Zhu et al. 2024).

Despite some progress, a persistent and overarching challenge is the fragmentation of regulatory development. In the absence of a globally harmonised legal framework, there is a risk of jurisdictions implementing incompatible standards, which will lead to regulatory arbitrage, higher compliance costs, and inefficiencies in operations. Furthermore, there is a risk that excessively strict or too imprecise legislation may discourage investments and developments, whilst an absence of regulation may result in unsafe or morally questionable deployments (Reddy et al. 2025).

To solve these issues, regulatory bodies must adopt adaptive governance, a paradigm that acknowledges the dynamism of technological development and includes mechanisms for iterative learning, stakeholder involvement, and cross-border harmonization. This includes not only rulemaking, but also the establishment of sandboxes, pilot projects, and international platforms for sharing lessons and co-developing standards. Like in the case of aviation and space law, the regulation of maritime autonomy will entail a long-term, globally inclusive process that will balance the need for advancement in technology with safety, accountability, and equity (Olaniyi et al. 2024).

3. ETHICAL CONSIDERATIONS

As the maritime industry transitions toward increased automation and autonomy, ethical considerations become a central, not peripheral, dimension of this technological evolution. Although the discussion on autonomous vessels and drones is largely couched in terms of operational efficiency and technical capability, these systems are intended to function in common, high-stakes situations in which their decisions have moral significance. Issues of responsibility, fairness, human dignity, and ecological integrity cannot be postponed to after-deployment audits; they must be part of the design, implementation, and governance of emerging maritime technologies.

One of the most pressing ethical challenges concerns accountability in autonomous decision-making. In traditional maritime operations, responsibility for decision-making rests with identifiable individuals: the shipmaster, the navigation crew, or the operator of a remotely piloted system. This chain of accountability is made diffuse with autonomous technologies. Who is liable morally and legally if an autonomous vessel harms – due to a navigational error, a software failure, or an unanticipated collision with another vessel? The outsourcing of decision-making to machines raises basic questions about moral agency and accountability, particularly in situations where lightning-fast decisions have to be made without human involvement (Lynch et al. 2024).

This issue is compounded by the challenge of algorithmic opacity. Many autonomous systems rely on complex artificial intelligence (AI) algorithms—particularly machine learning models—that are not easily interpretable, even by their developers. This "black box" problem leaves a hole in transparency that could erode trust and make accident investigations more difficult, and systemic biases in algorithmic training data more opaque. Explainability and transparency of maritime AI systems, especially those used in safety-critical operations, should thus be given priority by the ethical frameworks (Madsen, & Kim, 2024).

Another major ethical concern is the impact of automation on maritime labor. The global maritime workforce includes millions of seafarers, port workers, inspectors, and logistics personnel. The advent of autonomous ships and drones promises to replace much of these roles, especially in poor regions where maritime jobs are an economic mainstay. Although the proponents of automation tend to focus on job transformation instead of elimination, ethical deployment should address distributional justice, which means the burdens and benefits of innovation should be distributed fairly. This includes human capital investments such as reskilling the workforce, developing new system oversight and cybersecurity positions and including labor representatives in policy making (Sheikh ET AL. 2024).

The environmental ethics of autonomous maritime systems should also be closely examined. On the surface, automation appears to provide environmental benefits: streamlined routes can cut fuel use, unmanned vehicles can monitor ecological conditions with minimal disruption, and remote inspections may prevent pollution-causing accidents. However, more automation may allow for a rise in maritime activities, resulting in more cumulative ecological effects, particularly in sensitive marine regions. A precautionary approach is required for ethical deployment, which includes assessing long-term environmental trade-offs, incorporating marine conservation principles into system design, and aligning with the goals of the United Nations Sustainable Development Goals (SDGs), particularly those concerning climate action and life below water (De Beukelaer, 2024).

Additionally, the use of marine drones for surveillance and enforcement raises privacy and geopolitical concerns. Unmanned aerial and surface vehicles (UAVs) can monitor fishing activities, detect unlawful dumping, and enforce exclusion zones. Although such usage may contribute to environmental and legal compliance, they also risk increasing the state's monitoring powers, violating the privacy of maritime workers and small-scale fishers, and complicating standoffs in contested areas. Ethical governance must consider not just what these systems can achieve, but also who controls them, under what mandate, and how accountable they are (Islam, 2024).

The issue of ethical programming—the incorporation of moral decision-making frameworks into autonomous systems—is one of the most philosophically and practically hard. In high-risk scenarios, such as accident avoidance or emergency damage control, systems may be required to reconcile competing ethical ideals. Damage minimization, preserving lives, minimizing environmental damage, or adhering to legal standards. Following the lead of fields like as machine ethics and roboethics, marine developers will need to consider whether and how to program such principles into computers. Furthermore, the global diversity of legal, cultural, and philosophical ideas of ethics makes the standardization of such programming difficult (Sumartono, 2024).

Addressing these ethical concerns demands a multi-stakeholder, interdisciplinary approach. Ethical analysis cannot be an afterthought or the sole domain of developers. It must include ethicists, legal scholars, mariners, policymakers, and affected communities and civil society groups. Participatory design processes, public deliberation forums, and ethical audit mechanisms should be incorporated in all the stages of maritime innovation, including research and development, deployment, and decommissioning (Sepehri et al. 2024).

These ethical tensions are not perceived uniformly across the maritime ecosystem. Stakeholders — including ship owners, seafarers, port authorities, and regulators — differ significantly in how they view the risks and benefits of automation. While some anticipate operational gains, others highlight concerns about liability, employment, or infrastructure readiness. Table 1 summarizes these divergent viewpoints based on stakeholder group analysis from recent survey-based research.

Table 1. Summary of stakeholder perceptions regarding benefits and concerns of Maritime Autonomous Surface Ships (MASS).

Stakeholder Group	**Perception of Mass Benefits***	**Major Concern Highlighted***
Ship Owners / Operators	High (efficiency, cost savings)	Regulatory uncertainty, liability
Designers / Builders	Moderate-High	Technology readiness, integration complexity
Regulators & Flag States	Moderate	Legal / insurance frameworks, safety verification
Seafarers & Crews	Mixed	Displacement, changes to job roles, training demands
Port Authorities	Moderate	Infrastructure readiness, communication / coordination systems

In conclusion, the ethical dimension of maritime automation is not merely about avoiding harm or ticking compliance boxes. It is about confirming the values that the maritime sector - and society as a whole - wishes to maintain following its technological revolution. If ethics were placed at the heart of innovation, the maritime industry might navigate the complexities of autonomy with integrity, vision, and a shared purpose.

4. ECONOMIC DIMENSIONS

The economic ramifications of autonomous seafaring technologies are both exciting and disruptive. As shipping and maritime operations become more digitized and automated, a paradigm shift occurs, with far-reaching implications for cost structures, global trade dynamics, labor markets, and competitive posture. Although automation has the potential to increase efficiency and reduce long-term operating costs, it also introduces new layers of investment, maintenance, and systemic concerns, necessitating careful economic analysis.

At the heart of this transition is the promise of improved operational efficiency and cost savings. Autonomous ships and drones are frequently praised for their ability to reduce costs connected with crew salaries, accommodations, and safety infrastructure. For example, reduction or elimination of crew size enables redesign of the layout of vessels that can result in better space utilization and fuel economy. Automated systems also improve route optimization, idle times and human error – all of which lead to reduction in fuel consumption and insurance claims on collisions or mishandling of cargo (Riadh, 2024).

However, these efficiencies come at a significant initial cost. Developing, integrating, and certifying autonomous systems requires substantial capital expenditure. Costs consist of investments in high-precision navigation systems, satellite communications, remote operation centers, advanced sensors, AI algorithms, and cybersecurity infrastructure. Moreover, the companies have some costs for compliance with the changing regulatory standards, software updates, system redundancy, and staff training for remote monitoring and emergency intervention. These upfront financial barriers may be prohibitive to small and medium-sized enterprises (SMEs) and generates concern about market consolidation and access inequality in the nascent maritime economy (Liu et al. 2024).

The long-term economic sustainability of autonomous operations remains an open question. While there is focus on lifecycle savings and the possibility of predictive maintenance to minimize downtimes, lack of standardized business models and limited historical data make it hard to predict returns on investment with certainty. What is more, unforeseen expenses associated with system failure, cybersecurity breaches, or liability disputes can eat into expected profits. The insurance industry has specifically been circumspect on underwriting fully autonomous vessels due to uncertainties on risk assessment and claims processing (Diniz et al. 2024).

Beyond internal cost structures, automation is transforming maritime business models and logistics chains. The emergence of platform-based shipping solutions, drone-as-a-service offerings, AI-powered fleet management systems is opening new opportunities for value creation. These models can enable real-time cargo tracking, dynamic pricing, and an integrated port logistics, but they threaten traditional players to change or become obsolete. As digital intermediaries increasingly gain prominence, ports and shipping companies need to reimagine their place in increasingly data-driven supply chains (An, 2024).

The most far-reaching economic impact of automation may be felt in the labor market. Seafaring has historically provided employment to millions globally, particularly in lower- and middle-income countries. The gradual phase-out of aboard crewmen threatens to displace not only sailors but also a wide range of shore-side positions, including ship agents, dispatchers, trainers, and medical support personnel. Such disruptions have significant social and economic ramifications, particularly in maritime economies. As a result, morally and economically savvy policymakers will need to plan for and minimize employment losses through retraining, transition support, and the creation of new positions in system monitoring, analysis, and maintenance (Ji et al. 2024).

There is also a risk of a widening technological divide between countries and firms with the financial and technical resources to adopt automation and those without. Although technologically advanced countries may gain a competitive advantage through early adoption, less developed maritime economies may be left behind, sustaining global gaps in trade access, regulatory power, and value capture. This necessitates international collaboration in funding, capacity building, and technology transfer so that the benefits of marine innovation are distributed equitably (Song et al. 2025).

Furthermore, the macroeconomic consequences of significant automation in shipping must be considered. Autonomous operations may alter the frequency of shipping, cargo volumes, and port throughput, resulting in new requirements for port infrastructure, warehousing, and global supply chain design. Such systemic changes may alter trade routes, affect commodity prices, and influence national logistics policy (Asahi, 2024).

To summarize, the economic argument for maritime automation is clear, albeit complex. It is not a question of lowering expenses or raising profits. It necessitates an awareness of how evolving technologies alter economic relationships across the whole maritime value chain. Policymakers, investors, and industry leaders must view automation not only as an efficiency tool, but also as a transformative force that necessitates long-term planning, inclusive strategies, and flexible economic governance to maximize its benefits while managing its risks sensibly (Debelić, 2024).

5. INTERSECTIONS AND TRADE-OFFS

The challenges presented by emerging maritime technologies are rarely confined to isolated domains of regulation, ethics, or economics. Instead, they overlap in complex and conflicting ways to create trade-offs that require careful balancing on the part of stakeholders seeking to promote responsible and sustainable innovation. While maritime industry ventures into this technological frontier, it is becoming more evident that choices taken in one sphere – whether legal, moral, or financial – ricochet to the others, sometimes exacerbating the tensions and, at other times, opening opportunities for synergy.

A prominent intersection emerges between technological innovation and regulatory inertia. Autonomous systems are developing rapidly, often outpacing the capacity of legislative and institutional frameworks to respond. This temporal disjuncture results in what scholars term as "governance gap" wherein technology exists in a grey-zone of legality. Although early adopters can take advantage of the regulatory ambiguity to experiment with new systems, the lack of clear standards is a huge risk. These are legal uncertainty in case of accidents, inconsistency of enforcement in different jurisdictions, and proliferation of incompatible systems. The trade-off here is stark: or choke innovation by being too restrictive too early, or risk compromising public safety and legal accountability by moving too slowly (Kusumawati & Punomo, 2024).

As Figure 1 illustrates, the challenges associated with the adoption of Maritime Autonomous Surface Ships (MASS) are not restricted to technological maturity alone. A recent synthesis of the academic literature identifies regulatory and legal uncertainty as the most frequently cited barrier (35%), followed closely by technological limitations such as sensor reliability and system integration (30%). However, stakeholder resistance, cost-related barriers, and operational safety risks also feature prominently, highlighting the multidimensional nature of resistance to maritime automation. These barriers underscore the need for coordinated, cross-sectoral strategies that go beyond isolated technical fixes or piecemeal regulatory updates.

Figure 1. Relative frequency of barrier categories cited in recent literature on MASS adoption, based on thematic clustering of peer-reviewed studies

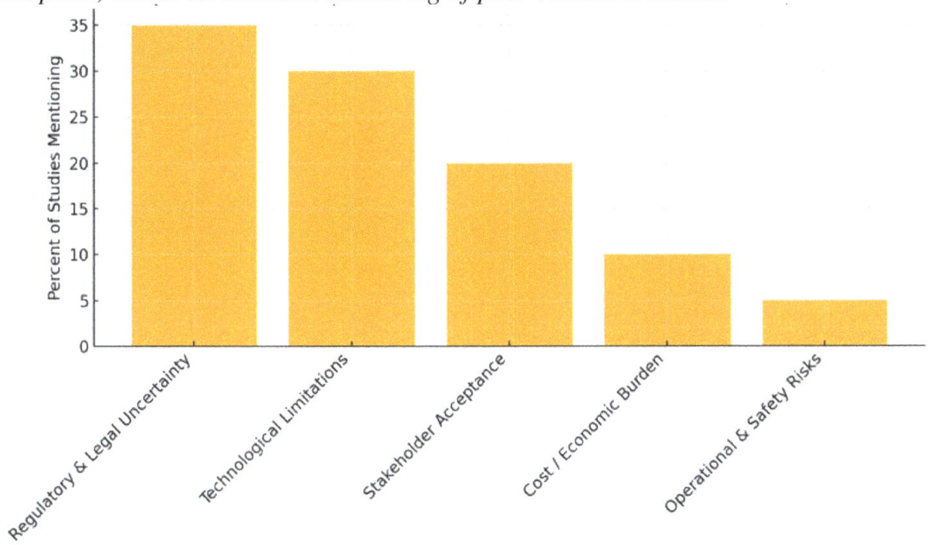

This dynamic is also reflected in the link between ethical precautions and economic imperatives. Incorporating strong ethical ideals into autonomous vessel design, such as justice, transparency, environmental responsibility, and data privacy, can improve trust, legitimacy, and long-term resilience. However, it may increase initial expenses, delay development timetables, or limit the flexibility of system architecture. For example, boosting explainability in AI decision-making may necessitate developers foregoing performance gains from more opaque machine learning models. Similarly, human intervention in critical situations can diminish the potential efficiency gains from complete automation. These trade-offs necessitate that developers and regulators establish a suitable balance between ethical rigor and market competitiveness (Karakas et al. 2024).

Surprisingly, ethics and economics are not always at opposed. Failure to implement ethical measures often has long-term economic effects. Adverse public reaction, tarnished reputation, and loss of stakeholder trust can sink projects or cause expensive legal liabilities. A case in point from other sectors is the "techlash" against unregulated digital platforms, and an outcome the maritime sector would do well to avoid. Preemptively adjusting technological systems to ethical norms and stakeholder expectations are increasingly becoming viewed as a risk management and value preservation tool, rather than an ethical responsibility (Okumus et al. 2024).

Another key point of contact is the effect of automation on labor and economic structures. While automation lowers operational costs and increases scalability, it also replaces human workers, creating ethical and economic concerns about regional inequality. This conflict is especially obvious in developing countries with a large seafaring population, where employment losses have the potential to destabilize entire communities. The trade-off here is the balancing of technological advancement and responsibility as a member of society. A profit-maximizing attitude may lead to socio-economic externalities which, if left unchecked, may trigger political opposition, regulatory counter-attacks, or demands for protectionism. A more inclusive strategy would invest in reskilling, build hybrid roles for human oversight, and support participatory transition planning that involves labor organizations (Autsadee et al. 2025).

Another trade-off scenario arises from the interconnectedness of environmental and economic objectives. Autonomous technologies offer promising tools for improving environmental performance—through optimized routing, predictive maintenance, and emissions reduction. However, the same technology can improve maritime activity by cutting marginal operating costs, hence increasing global shipping volumes and reducing the environmental burden on marine ecosystems. Regulatory regimes must thus be adjusted not only to encourage green technologies, but also to limit total environmental damage. Economic and environmental interests can coincide when incentives promote long-term sustainability over short-term exploitation (Karakas et al. 2024).

A new junction emerges in the sphere of international equity and technology access. High-income countries and significant shipping enterprises are most likely to take the lead in developing and implementing autonomous marine systems, which may exacerbate inequities in global maritime trade and technological governance. This raises ethical concerns about inclusivity and fairness, as well as economic issues like market concentration and dependency. A cooperative global system that includes technology sharing, capacity building, and financing for developing nations would not only reduce these disparities, but would also promote global maritime stability and the spread of innovations (Dominioni, 2024).

Finally, balancing these trade-offs highlights the importance of systems thinking. Conventional governance frameworks are likely to treat regulation, ethics, and economics in isolation, resulting in disjointed decisions and missed synergy opportunities. In contrast, systems thinking focuses on feedback loops, long-term effects, and inter-domain interactions. For example, a regulatory policy that incentivizes ethical AI design might reduce liability risk while increasing insurability, resulting in both ethical and economic benefits. Similarly, economic models that account for the externalities of labor displacement and environmental degradation are more likely to develop strong and adaptable corporate strategies (Garay-Rondero & Issa-Zadeh, 2025).

Overall, the future of marine automation will be defined not only by technological capabilities, but also by stakeholders' ability to navigate the complicated crossroads and compromises between innovation, governance, and justice. The challenge is to create systems and policies that do not simply maximize one dimension at the expense of another, but rather attempt to align them in an inclusive, adaptable, evidence-based manner.

6. RECOMMENDATIONS FOR POLICY AND PRACTICE

More than just technological readiness, the successful integration of emerging technologies into the marine domain necessitates coherent, inclusive, and flexible frameworks that match innovation with social values, legal standards, and economic stability. So far, regulatory, ethical, and economic analysis point to the need for actionable, multi-level strategies to guide stakeholders, such as governments, industry actors, technologists, civil society, and international institutions, toward a shared vision of responsible maritime autonomy. This section outlines essential recommendations for implementing such a vision.

6.1. Foster Global Regulatory Harmonization

To reduce uncertainty, avoid regulatory fragmentation, and ensure safety and interoperability, the maritime community must prioritize international regulatory harmonization. This involves speeding up the work of International Maritime Organization (IMO) to modernize and reinterpret the central conventions (for example, SOLAS, COLREGs, and STCW) in the context of autonomous technologies. Stakeholders should lobby for clear definitions of the levels of autonomy, standardized protocols for remote operation, and uniform guidelines for vessel registration, liability, and insurances. Global testbeds with shared data repositories could enable collaborative regulatory development while also allowing for country variances (Koo et al. 2024).

Regional groups (such as the EU, ASEAN, and the African Union) can serve as bridges, testing unified frameworks and growing capacity in member states with limited resources. These initiatives should be built on the values of openness, safety, and technological neutrality in order to foster innovation within a predictable and equitable legal framework.

6.2. Embed Ethical Governance in System Design

Ethical considerations must be integrated into all stages of technology development and deployment, not just as afterthoughts. This necessitates the formation of multi-stakeholder ethical review committees for maritime AI systems, comprising ethicists, sailors, legal experts, community representatives, and technologists. These boards should evaluate systems based on criteria like as transparency, accountability, explainability, environmental effect, and fairness, particularly in terms of labor and data privacy.

Developers must adopt ethics-by-design thinking, which incorporates principles such as human dignity, environmental sustainability, and social inclusion into system architectures and operational protocols. Standards organizations and classification societies can incorporate ethical benchmarks into certification requirements, making ethical performance a competitive benefit rather than a compliance cost (Zhu et al. 2024).

6.3. Advance Inclusive Economic Transition Strategies

To ensure that economic benefits are equitably distributed and societal harms are mitigated, governments and industry must implement inclusive transition strategies. This involves investing in reskilling and upskilling of displaced maritime workers through public-private training programs, which focuses on digital navigation, AI oversight, system diagnostics, and maritime cybersecurity. The focus should be paid to workers in developing regions, where the economic and social effects of automation can be more pronounced.

Policy tools like tax incentives, innovation grants and social safety nets can help small and medium sized enterprises (SMEs) and vulnerable labor markets. On the international scene, development banks and maritime organizations should establish funding mechanisms and agreements for transfer of technology to make sure that the poor countries do not lag behind in the autonomous transition (Elston et al. 2024).

6.4. Promote Data Governance and Cybersecurity Standards

As maritime systems become increasingly reliant on digital infrastructure, robust data governance and cybersecurity policies are critical. National and international maritime regulators should collaborate with standards bodies (e.g., ISO, IEC) to create enforcable guidelines for data protection, integrity, access control, and system resilience. These guidelines should take into account the specific vulnerabilities of autonomous systems, such as sensor spoofing, communication jamming, and software corruption.

Operators should be required to maintain audit trails, obtain cybersecurity certification, and use real-time threat detection. Furthermore, the ethical utilization of surveillance data, particularly that collected by drones used in law enforcement or environmental monitoring, should adhere to strict privacy rules and be independently verified (Inal, 2024).

6.5. Institutionalize Cross-Sector Collaboration and Foresight

The complexity of autonomous maritime systems needs constant cross-sectoral collaboration. National and regional maritime innovation councils or public-private advisory boards should be established, with legislators, technologists, labor unions, environmental groups, insurers, and academic researchers serving on them. These platforms can act as early warning systems for emerging hazards, as well as promote trust and policy innovation.

Furthermore, governments should promote scenario planning and foresight exercises that take into account the long-term consequences of maritime automation in various regulatory frameworks, environmental, and geopolitical contexts. This forward-thinking governance strategy can assist stakeholders in anticipating challenges, increasing resource efficiency, and developing contingency plans prior to disasters (Palmer-Abbs et al. 2024).

6.6. Integrate Innovation with Environmental and Social Goals

Finally, the transition to marine autonomy must be consistent with broader goals for justice and sustainability. Autonomous technologies should be judged not only on their technical prowess and financial return, but also on how successfully they promote social inclusion, marine biodiversity protection, and climate action. Private incentives can be combined with public goods by utilizing policy instruments such as sustainability-linked subsidies, carbon credits for autonomous ships with low emissions, and environmental impact assessments.

The inventors should be encouraged to take a systems-level approach, recognizing that the health of the oceans, the resilience of coastal inhabitants, and the strength of governing institutions all influence how sustainable marine technology is.

Finally, the wise use of autonomous maritime systems will not be driven solely by market forces or technological talent. It necessitates intentional, forward-thinking, inter-sectoral, and cross-border action. Following the ideas made here, the maritime community can ensure that the digital transformation of the seas is not only effective and creative, but also fair, ethical, and sustainable (Diniz et al. 2024).

7. LIMITATIONS AND SCOPE BOUNDARIES

While this chapter presents an interdisciplinary framework for assessing regulatory, ethical, and economic dimensions of maritime automation, several limitations must be acknowledged:

1. Conceptual Over Empirical Approach

The chapter offers a systems-level conceptual synthesis rather than empirical case studies. While this allows for broad applicability, it limits specificity in evaluating how particular technologies perform in different maritime contexts.

2. Evolving Technological Landscape

The field of maritime autonomy is evolving rapidly. Many regulatory or technical developments cited here may change in short time frames, and the chapter does not claim to capture a static or exhaustive state of the field.

3. Geographic and Policy Bias

Much of the literature and regulatory progress discussed originates from high-income maritime nations (e.g., Norway, Singapore, EU), which may not reflect the realities or constraints in developing economies.

4. Ethical Variability

Ethical analysis is context-sensitive and culturally contingent. The frameworks proposed here are grounded in universalist principles, but real-world adoption may vary depending on legal traditions, political will, and societal values.

5. Economic Assumptions

Forecasts of cost-efficiency, labor displacement, or market disruptions are based on existing projections, not longitudinal evidence. The actual economic impact of automation will depend on many unknown variables, including future trade volumes, insurance policies, and public perception.

Recognizing these limitations is essential for interpreting the findings of this chapter as directional rather than definitive. Future work should test and refine the ideas proposed here through empirical field studies, comparative legal analyses, and stakeholder-driven ethical impact assessments.

8. CONCLUSION

The maritime sector is undergoing a technological transformation driven by autonomous ships, drones, and digital systems. This chapter has shown that while these innovations promise gains in efficiency, safety, and environmental performance, they also disrupt existing regulatory norms, ethical responsibilities, and economic structures.

Key findings from our analysis include:

- Regulatory gaps persist due to outdated conventions and lack of harmonized global frameworks, particularly in vessel flagging, liability, and cybersecurity.
- Ethical challenges such as algorithmic opacity, labor displacement, and surveillance risks require value-sensitive design and inclusive governance models.
- Economic transformations are unevenly distributed—automation offers operational gains but imposes high entry costs and exacerbates global inequalities in access to maritime innovation.

To ensure maritime autonomy is responsible and sustainable, stakeholders should:

- Accelerate international regulatory alignment (e.g., through IMO conventions and regional testbeds).
- Institutionalize ethics-by-design frameworks with multi-stakeholder oversight.
- Support inclusive economic transitions, especially for seafarers and developing economies.
- Invest in cybersecurity and data governance as core pillars of system resilience.
- Integrate environmental and social goals into technical innovation metrics.

Future work should prioritize longitudinal studies on the socio-economic impacts of automation, development of ethical audit tools for maritime AI, and adaptive policy experiments in real-world deployment contexts.

REFERENCES

Ahmed, Y. A., Theotokatos, G., Maslov, I., Wennersberg, L. A. L., & Nesheim, D. A. (2024). Regulatory and legal frameworks recommendations for short sea shipping maritime autonomous surface ships. *Marine Policy*, *166*, 106226. DOI: 10.1016/j.marpol.2024.106226

An, J. (2024). Maritime logistics and digital transformation with big data: Review and research trend. *Maritime Business Review*, *9*(3), 229–242. DOI: 10.1108/MABR-10-2023-0069

Animah, I. (2024). Application of bayesian network in the maritime industry: Comprehensive literature review. *Ocean Engineering*, *302*, 117610. DOI: 10.1016/j.oceaneng.2024.117610

Asahi, S. (2024). Economic Ripple Effects of Choice Factors and Disaster-Related Investments at Nagoya Port. In *Econometric Analysis of Regional Economy* (pp. 185–230). Springer Nature Singapore. DOI: 10.1007/978-981-99-8707-8_5

Autsadee, Y., Jeevan, J., Bin Othman, M. R., & Mohd Salleh, N. H. B. (2025). Maritime Society 5.0: A global transition on human economy and civilisation for maritime sustainability. *Australian Journal of Maritime and Ocean Affairs*, *17*(1), 1–26. DOI: 10.1080/18366503.2023.2287872

De Beukelaer, C. (2024). Shipping the future: Climate ethics for maritime transport. *ephemera: theory & politics in organization, 24*(1).

Debelić, B. (2024). Economic Governance in Maritime Affairs. *Pomorski zbornik, 64*(1), 135-156.

Diniz, N. V., Cunha, D. R., de Santana Porte, M., Oliveira, C. B. M., & de Freitas Fernandes, F. (2024). A bibliometric analysis of sustainable development goals in the maritime industry and port sector. *Regional Studies in Marine Science*, *69*, 103319. DOI: 10.1016/j.rsma.2023.103319

Dominioni, G. (2024). Carbon pricing for international shipping, equity, and WTO law. *Review of European, Comparative & International Environmental Law*, *33*(1), 19–30. DOI: 10.1111/reel.12540

Elston, J., Pinto, H., & Nogueira, C. (2024). Tides of Change for a Sustainable Blue Economy: A Systematic Literature Review of Innovation in Maritime Activities. *Sustainability (Basel)*, *16*(24), 11141. DOI: 10.3390/su162411141

Farah, M. B., Ahmed, Y., Mahmoud, H., Shah, S. A., Al-Kadri, M. O., Taramonli, S., Bellekens, X., Abozariba, R., Idrissi, M., & Aneiba, A. (2024). A survey on blockchain technology in the maritime industry: Challenges and future perspectives. *Future Generation Computer Systems*, *157*, 618–637. DOI: 10.1016/j.future.2024.03.046

Garay-Rondero, C. L., & Issa-Zadeh, S. B. (2025). Integrating sustainability thinking in maritime education. *Cogent Education*, *12*(1), 2479205. DOI: 10.1080/2331186X.2025.2479205

Hebbar, A. A. (2024). Implementation of IMO instruments from a flag state perspective. *The Elgar Companion to the Law and Practice of the International Maritime Organization*, 304-329.

Inal, O. B. (2024). Decarbonization of shipping: Hydrogen and fuel cells legislation in the maritime industry. *Brodogradnja*, *75*(2), 1–13. DOI: 10.21278/brod75205

Islam, M. S. (2024). Legal implications of emerging technologies in maritime pollution monitoring and management. *Journal of the Indian Ocean Region*, *20*(1), 54–75. DOI: 10.1080/19480881.2024.2387819

Ji, J., Chi, Y., & Yin, X. (2024). Research on the driving effect of marine economy on the high-quality development of regional economy–Evidence from China's coastal areas. *Regional Studies in Marine Science*, *74*, 103550. DOI: 10.1016/j.rsma.2024.103550

Karakas, S., Acar, A. Z., & Kirmizi, M. (2024). Maritime Sustainability: Navigating Complex Challenges and Ecological Footprints. In *Sustainable Development Seen Through the Lenses of Ethnoeconomics and the Circular Economy* (pp. 19–47). Springer Nature Switzerland. DOI: 10.1007/978-3-031-72676-7_2

Koo, K. Y., Rødseth, Ø. J., Lislebø, G., & Ulvensøen, J. H. (2024, October). Harmonizing Maritime Innovation: Enhancing International and National Standardization in Intelligent Ship Transport Systems. [). IOP Publishing.]. *Journal of Physics: Conference Series*, *2867*(1), 012023. DOI: 10.1088/1742-6596/2867/1/012023

Kusumawati, E. D., & Punomo, B. (2024). Technological Transformation and Innovation Strategy as a Key Pillar in Improving Sustainability Efficiency In The Maritime Industry. *IITE Proceeding: International Inovation Technology Proceeding, 2*(1), 65-76.

Liu, J., Liao, R., Dong, F., Huang, C., Li, H., Liu, J., & Zhao, T. (2024). Low-carbon technology selection and carbon reduction potential assessment in the shipbuilding industry with dynamically changing grid emission factors. *Journal of Cleaner Production*, *441*, 140707. DOI: 10.1016/j.jclepro.2024.140707

Lynch, K. M., Banks, V. A., Roberts, A. P., Radcliffe, S., & Plant, K. L. (2024). What factors may influence decision-making in the operation of Maritime autonomous surface ships? A systematic review. *Theoretical Issues in Ergonomics Science*, *25*(1), 98–142. DOI: 10.1080/1463922X.2022.2152900

Madsen, A. N., & Kim, T. E. (2024). A state-of-the-art review of AI decision transparency for autonomous shipping. *Journal of International Maritime Safety, Environmental Affairs, and Shipping*, *8*(1-2), 2336751. DOI: 10.1080/25725084.2024.2336751

Okumus, D., Gunbeyaz, S. A., Kurt, R. E., & Turan, O. (2024). An approach to advance circular practices in the maritime industry through a database as a bridging solution. *Sustainability (Basel)*, *16*(1), 453. DOI: 10.3390/su16010453

Olaniyi, E. O., Solarte-Vasquez, M. C., & Inkinen, T. (2024). Smart regulations in maritime governance: Efficacy, gaps, and stakeholder perspectives. *Marine Pollution Bulletin*, *202*, 116341. DOI: 10.1016/j.marpolbul.2024.116341 PMID: 38608427

Palmer-Abbs, M., Deshpande, P., & Karl, C. W. (2024). Enabling transition thinking on complex issues (wicked problems): A framework for future circular economic transitions of plastic management in the Norwegian fisheries and aquaculture sectors. *Journal of Cleaner Production*, *449*, 141420. DOI: 10.1016/j.jclepro.2024.141420

Pareek, P., Hajra, R., & Mohapatra, H. (2025). Empowering Public Officials by Bridging the Digital Divide with Smart Technology for Public Services. *Digital Competency Development for Public Officials: Adapting New Technologies in Public Services*, 419-452.

Potamos, G., Stavrou, E., & Stavrou, S. (2024). Enhancing maritime cybersecurity through operational technology sensor data fusion: A comprehensive survey and analysis. *Sensors (Basel)*, *24*(11), 3458. DOI: 10.3390/s24113458 PMID: 38894249

Reddy, A. S., Ashraf, M., Manideepak, T., & Mohapatra, H. (2025). A comprehensive analysis of governance and regulatory frameworks for cloud-based initiatives for smart cities. In *Organizational sociology in the digital age* (pp. 127–146). IGI Global Scientific Publishing. DOI: 10.4018/979-8-3693-7398-9.ch008

Riadh, M. (2024). Evaluating the environmental and economic impacts of MARPOL compliance in the global shipping industry. *Collaborate Engineering Daily Book Series*, *2*(2), 98–104. DOI: 10.62012/collaborate.v2i2.70

Sepehri, A., Kirichek, A., van den Heuvel, M., & van Koningsveld, M. (2024). Smart, sustainable, and circular port maintenance: A comprehensive framework and multi-stakeholder approach. *Journal of Environmental Management*, *370*, 122625. DOI: 10.1016/j.jenvman.2024.122625 PMID: 39316882

Sheikh, W., Islam, T. R., & Islam, H. (2024). Analyzing the impact of organizational justice on seafarers' commitment: Job satisfaction as a key mediator. *Australian Journal of Maritime and Ocean Affairs*, •••, 1–31. DOI: 10.1080/18366503.2024.2403741

Song, Z. Y., Chhetri, P., Ye, G., & Tae-Woo Lee, P. (2025). Green maritime logistics coalition by green shipping corridors: A new paradigm for the decarbonisation of the maritime industry. *International Journal of Logistics*, *28*(4), 363–379. DOI: 10.1080/13675567.2023.2256243

Sumartono, E. (2024). The Life of Seafarers in the Digital Age: An Analysis of the Philosophy of Ethics and Morality in the Use of Technology. *Side. Scientific Development Journal*, *1*(3), 78–87.

Voroniuk, N. (2024). Relevance of international legal and national standards of safety of navigation and the role of classification societies in the context of modern challenges and innovations in shipbuilding. *Scientific Journal of the National Academy of Internal Affairs*, *4*(29), 22–34. DOI: 10.56215/naia-herald/4.2024.22

Wang, X., Guo, Y., & Gao, Y. (2024). Unmanned autonomous intelligent system in 6G non-terrestrial network. *Information (Basel)*, *15*(1), 38. DOI: 10.3390/info15010038

Zhu, M., Zhang, W., & Xu, C. (2024). Ethical governance and implementation paths for global marine science data sharing. *Frontiers in Marine Science*, *11*, 1421252. DOI: 10.3389/fmars.2024.1421252

Zhu, Q., Xi, Y., Weng, J., Han, B., Hu, S., & Ge, Y. E. (2024). Intelligent ship collision avoidance in maritime field: A bibliometric and systematic review. *Expert Systems with Applications*, *252*, 124148. DOI: 10.1016/j.eswa.2024.124148

Chapter 6
Role of Fighter Drones in Maritime Applications

G. Prasad
https://orcid.org/0000-0002-5709-9182
Chandigarh University, Punjab, India

Snehal Malik
https://orcid.org/0009-0006-0823-1446
Chandigarh University, Punjab, India

Aadya Gupta
https://orcid.org/0009-0005-3149-6584
Chandigarh University, Punjab, India

ABSTRACT

This chapter will consider the transformations brought about in functions performed by military-grade unmanned aerial vehicles (UAVs)-that is, fighter drones-as it concerns their role in contemporary maritime operations and naval warfare. Networked communications, sensor technologies, autonomy, and artificial intelligence (AI) are paving the way in the new opportunities to fill the area over which UAVs will become permanent materials in the strike missions, anti-submarine warfare, search and rescue, and environmental monitoring which can, prior, be only covered by ISR. This chapter also discusses the literature on core theory and current operations practice, the technology behind UAV effects in adverse marine environments, and capabilities and functions with which these technologies are associated.

DOI: 10.4018/979-8-3373-3546-9.ch006

INTRODUCTION

Once remote drone technologies in naval warfare were constantly overshadowed by alternative technologies, remote pilots have become core indispensable platforms of the revolutionary equal manner development and with maritime missions perpetually their persistence, enhanced precision, and flexibility in accomplishing them. Besides ISR, strikes, and the monitoring of environmental conditions, rescue activities are also assisted by the search deployed fighter drones which can be found to perform autonomously or semi-autonomously, thus making them essential to the progression of the naval battlespace (Ruiz Estrada, 2024). As illustrated in this chapter, the options of the implementation of their application include gray zone conflict management, deep-sea surveillance, swarm-based combat, strategic deterrence, etc. Their offensive and defensive capabilities will very much increase the level of naval warfare; however, the integration of fighter drones also brings complications, such as, hiccups in the level of endurance, failure in communication systems, hacking in GNSS-denied environments, bad weather-related issues, lack of trained personnel even to ensure the design, deployment, and maintenance of drones. Besides that, ethical concerns on open-lethal autonomous systems, escalation risks, and the dilution of accountability over unmanned engagements that only make difficult the deployment of these drones in the waters where fights take place should also be considered. Overcoming these obstacles, the technology will finally empower these drones to have enhanced intelligence, survivability, and strategic significance. 6G connection, quantum navigation, AI swarm capabilities, modular payloads, and underwater aerial drone interoperability will be the factors that will bring a revolution to naval warfare. Besides that, with logistics and maintenance systems becoming smarter and more autonomous, there will be long-term benefits of costs and maintenance for sustained presence and dispersed operations of drone fleets.

Fundamentally, fighter drones would be the primary form of the future strategy, doctrine, and outcome of naval conflicts when the navies worldwide adapt to a maritime environment that is increasingly contested. The moral, legal, and strategic implications of drone warfare would require corresponding increments in policy, regulation, and international norms to ensure that their use is both efficient and accountable.

BACKGROUND OF THE RESEARCH

This review of literature describes the transformation of the use of drone (Unmanned Aerial Vehicle or UAV) in the sea environment with an emphasis on their potential, difficulties, and their possibilities in the future. The review presents a holistic approach by explaining the range and importance of unmanned aerial vehicles (UAVs) in the maritime environment and subsequently, bringing up the issues, the gaps in research, and finally, summarizing the findings and the future research openings. In the present-day scenario, UAVs are gradually becoming an indispensable part of routine maritime operations, thus, offering the capability of constant surveillance, ISR, and intelligence. The complete range of UAVs for maritime purposes can be visualized from patrolling and law enforcement to search and rescue and environmental observations (Frederiksen & Knudsen,2018). A few of the reports that emphasize some of the numerous functions of a UAV include monitoring of sulfur emissions on board ships, antenna testing onboard ships, and inspection of offshore wind farms (Frederiksen & Knudsen,2018) and comprehensive overviews of sensor and platform technologies (Wang et al., 2023). The debate of the relevance of the military application of UAV-likewise is going on so intensely that topics like the types of UAV, their application in reconnaissance and surveillance, and their research in countries like India and Israel (Duan & Zhang, 2014) are discussed.

Comparative analyses of global UAV platforms further highlight capability gaps (Singh, 2016). The US Navy, specifically, has a multi-level UAV integration plan, with a concentration on tactical UAVs to complement battle space management and enhance situational awareness ((Nomikos et al., 2022). Furthermore, Recent studies emphasize the importance of BVLOS (Beyond Visual Line of Sight) operations for naval UAVs, especially in the context of secure, long-distance command and control in areas needing significant surveillance.

In the past previous two decades, the fighter drone development for mobile strikes and reconnaissance over the seas has been developed to be one of the extreme innovations ever witnessed in naval warfare, along with the technology for autonomous systems, artificial intelligence, and communications in support of it. The inter-disciplines, while having intensified unequally in areas of research, delved further into the maritime applications dictated by the uses of UAVs in maritime surveillance, combat, and maritime support.

According to Taylor (2021), work on counter-swarm defense matrix against shipboard lasers indicated that diverse drone swarms represented a heterogeneous problem and so deserved an agile engagement solution. In simulation-supported analyses, he pointed out that besides the behavior of swarm drones, the sequence of attacks has significant implications for the success of engagements, especially under contested maritime environments. These are the same studies that were done concurrently between 2021 and 2024, focusing on the cyber resilience of the maritime sector as well as on the safeguards against GNSS jamming/spoofing.(Nomikos et al., 2022) presented integrated techniques combining INS, visual odometry, and AI-assisted navigation for the regions where GNSS signals have been blocked while

(Nomikos et al.,2022) investigated the safety of drone-based communication relays facing the risk of spoofing.

Surveillance and threat detection are one major area for applications at sea for UAVs. Wang et al., (2023) discussed the fast-growing application of UAVs with AI-assisted real-time object recognition for the maritime domain. These UAVs are capable of detecting illegal activities, such as fishing or oil spills, and emergency incidents, even in a faster and more precise manner. Ramesh & Jeyan, (2020) have taken this work to the next level by comparing the performance of deep learning-based unmanned aerial vehicle (UAV) detection in maritime search and rescue (SAR) operations under different sea conditions. Their results showed that deep learning gave them considerable performance improvements. One of the latest studies by Wang et al., (2023) shows that the AI-based drone can be trusted to perform in real-life SAR missions because the technology can achieve real-time object recognition with high accuracy even in turbulent waters. This system is a tremendous development, allowing great enhancement in the reactivity of search-and-rescue (SAR) operations, especially when shipwrecks or emergencies arise overboard.

Despite the expanding horizon for UAVs, there are many challenges and unfulfilled research opportunities. Regulatory issues, particularly regarding the capability for technical rules regulation for Beyond Visual Line of Sight (BVLOS) operations, are major hindrance to full realization (Frederiksen & Knudsen,2018). Technological matters, including battery life, communication systems, and failsafe systems, are other areas that will benefit from further improvement (Leman & Nevalainen, 2025). UAV integration into current maritime doctrine and demand for enunciated road map for tactical UAVs are also areas identified for enhancement (Nomikos et al., 2022). Indigenous development of UAVs in countries such as India also faces challenges due to critical technology issues and attracting talent (Leman & Nevalainen, 2025).

INTERNATIONAL COLLABORATION AND INTEROPERABILITY STANDARDS

Usually, multinational collaboration is required for maritime operations, and such a partnership implies the existence of communication and data standards that would be common for all parties. One of the International Hydrographic Organization (IHO) standards, the S-100 framework is a set of interoperable maritime data models that enable uniform mapping and exchange of information between different navies. Communication technologies like MQTT and OPC UA are gaining more popularity in maritime IoT as they allow secure and low-latency data transfers between drones, ships, and command centers. Hence, by following these standards, allied fleets will not only be able to use different types of drones from various manufacturers, but also perform better-united operations thus, they can increase not only coalition effectiveness but also interoperability during joint missions.

To be sure, UAVs are reshaping the way naval units conduct their missions by providing them with additional ISR and operational flexibility. They have achieved a good deal of progress; however, the settling of regulatory and technological issues is what will really pave the way for the unfolding of their full potentials. Subsequent studies should aim at building robust BVLOS capabilities, promoting AI integration, and generating more detailed strategic roadmaps for the use of tactical UAVs to deliver a complete picture of the Maritime Domain. The expansion of a fleet of air and underwater unmanned vehicles is a radical change in the capabilities of naval warfare of the near future. Most of the pieces of evidence examined show a definite trend: naval combat drones have evolved from being only ISR platforms to becoming highly capable, autonomous, multi-role, intricately ship network-connected systems. Their ability to operate in bad weather, denied areas, and contested regions makes them extremely valuable for the present maritime strategy.

TECHNOLOGICAL FOUNDATION OF MARITIME FIGHTER DRONES

Solo naval-beating drones is almost entirely a scenario of advanced tech in which all have to be put to work for autonomy, range of long endurance, the capacity of many other kinds of missions, and the ability to integrate the command systems of the navy. Besides, these technologies are a host of subsystems ranging from high-speed sensors to fault-tolerant communication links and AI-based autonomy and swarm coordination capability (Kumar & Sharma, 2022). The sensor arrays are the principal weapons through which the UAVs grasp and operate in the maritime environment. The unmanned combat vehicles employ a combination of electro-optical (EO) and infrared (IR) sensors for visual acquisition, day/night reconnaissance, and target detection. The sensors provide high-resolution pictures and thermal traces that are very useful for ISR, targeting, and battle damage assessment.

Besides, multispectral and hyperspectral imaging technologies have gone further to extend UAV capabilities by the ability to perceive spectral signatures—very important during the detection of oil spills, hidden ships, and human bodies in SAR operations. For instance, Duan & Zhang, (2014)utilized multispectral UAVs for the speedy detection of drowning victims in marine rescue operations through neural networks trained from spectral data.

LiDAR (Light Detection and Ranging) is a prominent sensor system, especially for low-altitude UAVs in the vicinity of coastlines, naval installations, or sea surface assets. Novel capsule-based sensor pods have also been trialed for hull inspections

(Nomikos et al., 2022). It allows for 3D mapping, high-accuracy altimetry, and obstacle avoidance, which is especially beneficial in autonomous landing systems on traveling ships or drone carriers. With the advancement of sensor fusion techniques, existing fighter drones increasingly integrate data from various sources (EO/IR, LiDAR, radar) to yield a unified situational picture.

1. Data Links and Communications

There should be secure, high-bandwidth communication to support command, control, and data interchange between UAVs and warships. Secure line-of-sight (LOS) radio frequency (RF) communications are employed by fighter UAVs for tactical, short-range missions and Beyond Visual Line of Sight (BVLOS) links—often by satellite or high-altitude relay UAVs—for enduring operations.

BVLOS capability is critical in naval applications where ships might be hundreds of kilometers from shore or supply stations. The issue is to provide low-latency, high-integrity, jam-resistant communication, especially in hostile or GNSS-denied areas.

Frequency hopping, encryption algorithms, software-defined radio (SDR), and directional antennas have been implemented on maritime drones to counter them.

According to Nomikos et al., 2022), UAV use is increasingly relevant in maritime communication networks, especially for maritime IoT, SAR missions, and real-time ISR applications. UAVs not only operate as data gatherers but also as airborne nodes to provide communication extensions between surface ships, submarines, and command-and-control centers. The recent surveys underscore design conflicts over link deployment, channel modeling, and latency management for maritime UAV networks (Gans & Rogers, 2021).

2. Autonomy and AI-Based Decision Making

From simple remote-controlled aircraft, UAVs have now become highly autonomous agents that are able to make real-time decisions by deploying artificial intelligence (AI) within them. AI algorithms allow navy drones to detect and classify threats and targets, avoid obstacles, and plan missions based on situational changes without human involvement.

Employed mission control is particularly effective for high-risk or challenging time-critical actions where human involvement can increase delay. For example, convolutional neural networks (CNNs) are applied to detect objects such as ships, lifeboats, and submarines from multispectral imagery. Reinforcement learning and fuzzy logic controllers are being investigated for not only incorporating these approaches into UAV flight in maritime use but also allowing the participant to determine the appropriate flight paths, loiter pattern, and methods of engagement.

3. Swarm Intelligence and Combat Coordination

One of the most revolutionary developments in fighter drone warfare is the creation of drone swarms—groups of UAVs that operate based on the principles of swarm intelligence. Inspired by nature through studying flocks of birds or schools of fish, swarm algorithms allow for decentralized self-organization of drones The U.S. DARPA OFFSET program has explored many of these swarm tactics in littoral missions (DARPA, 2021).

Several recent naval research works have checked how effective the shipboard Laser Weapon Systems (LWS) are to the coordinated UAV swarm threats. Moshiri, Azadi and Malek (2020) mentioned the difficulties which a laser-guided energy weapon could face when used against a drone swarm, whereas the simulation experiments of the U.S. Navy in 2022–2023 have revealed that laser defenses could still be overwhelmed by the adaptive drone formations under turbulent sea conditions. His simulations revealed that swarm attack effectiveness is largely dependent on formation, direction of attacks, and timing—conditions that can be optimized using AI and environmental feedback in real-time. Ruiz Estrada, (2024)

Swarm coordination also includes cooperative participation, in which numerous drones share sensor data and processing power to pursue and attack moving targets, such as high-speed attack boats or sub-surface-penetrating submarines. Swarms can also execute decoy operations, whereby UAVs are sacrificed as decoys to attract ship defenses away while others gather intelligence or release ordnance.

To enable such coordination, novel protocols like Vehicle-to-Vehicle (V2V) communication, mesh networks, and blockchain-based command integrity are being studied for trustworthy, scalable swarm operation in naval environments.

Figure 1. Coordinated Sensor Coverage and Communication in a UAV Swarm Network (Gans & Rogers, (2021)).

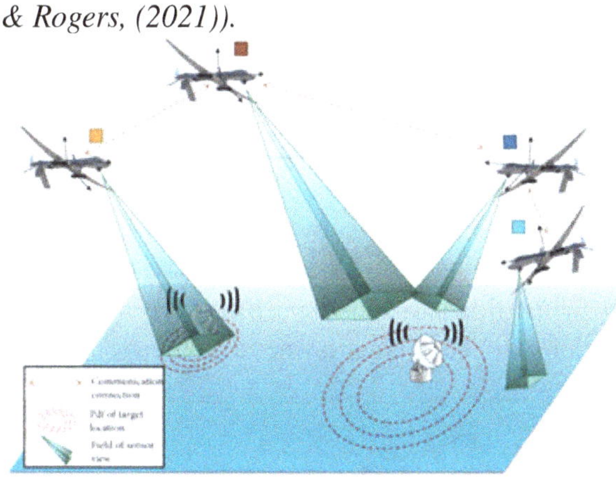

The dual mission of the synchronized UAV swarm depicted in Figure 1 is monitoring over maritime areas and target-following activities. The exo-drones continuously monitor the sensor coverage over some area and communicate with each other over secure links (denoted with red dashed lines) and somehow overlap in fields of view and data sharing, but allow for accurate and agile cooperation in tracking likely target locations (depicted as red ellipses)—a real example of the principles of distributed sensing and cooperative engagement in modern naval operations.

Essentially, the technology behind the fighter drones that are launched from the sea is quite complex and keeps on getting better at a fast pace. This progress is highly driven by the demands for high survivability, mission autonomy, and extended range in the maritime battle space. From hardened communications and advanced sensors to AI autonomy and swarm logic, these technologies are reshaping the future of naval power projection. Being able to see and invest in such root technologies is vital for navies that wish to gain ascendancy in the emerging unmanned and networked maritime theater.

TYPES AND CLASSIFICATIONS OF MARITIME FIGHTER DRONES

Figure 2. Classification of UAV based on wings and rotors (Ramesh & Jeyan, 2020)

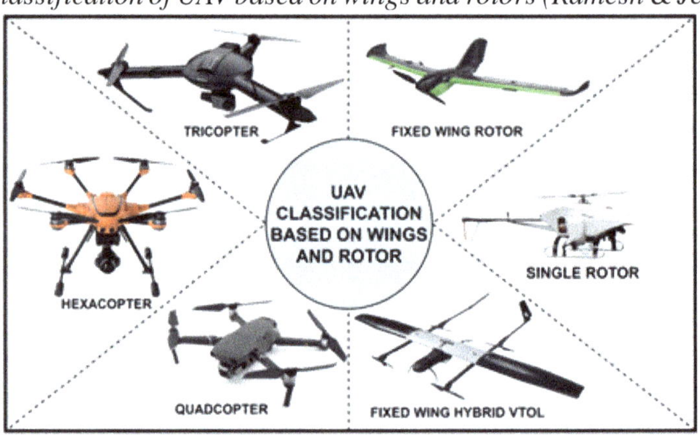

Figure 2 maps the taxonomy of UAV platforms based on their wing and rotor configurations, which form the basis of their performance in maritime operations. The following configurations include fighter drones designed for maritime applications:

1. Fixed-Wing UAVs

Being winged aircraft, these are rather the most commonly used ones in maritime settings with respect to time-in-air and flight ranges. The stiff wings generate lift through forward motion; hence, they can sustain even endurance missions, including wide-area ocean surveillance, intelligence gathering, and early threat warning.

Forms of fixed-wing flight depend upon the altitudes, operating mostly in medium to high elevations and being categorized by the payload, such as radar, EO/IR cameras, and other electronic warfare systems. A few hours can scan a wide area of interest over water. This is even more so when looking into the large expanse of ocean known as Exclusive Economic Zones (EEZ) or reconnaissance along the highly contested sea routes.

2. Rotary-Wing UAVs

VTOL helicopters-and multi-rotor UAVs are some of the rotary-wing UAVs that are fit for operations from ships experiencing limited spaces. Such remotely controlled flying vehicles are really brilliant in tactical roles, for instance, close-in surveillance, SAR(Search and Rescue), ship inspections, and target tracking.

The most important thing is that these devices can keep a hover-position and can follow the movement very accurately, thus they are able to perform detailed observation tasks, they can get into or out of the cluttered places, and can be of help to the boarding team or the maritime patrol. Rotary UAVs, which can be launched from various assets, such as destroyers, frigates, and coast guard cutters, are free from the necessity of runways or any other launching systems. These helicopters show a lesser flying time and range.

3. Tactical UAVs

UAVs Such kind of UAVs is aimed to be developed for short-range high-tempo missions and real-time data providing along with an alerting feature. The design is compact and the weight is low enough so that they could be carried by a man without the need of additional equipment and thus be easily launched directly from naval vessels or even

small platforms without the necessity of expensive infrastructures. They are assigned to conduct ISR missions at the fleet or ship level to give commanders the real-time situation awareness against an ever-changing operational background. Typically used for target acquisition, battle damage assessment, electronic support, and maritime interdiction.

4. Strategic UAVs

These are offshore, high-endurance platforms able to picture massive maritime zones for protracted periods. Such platforms equipped with a suite of sensors and secure communication links, at times even offensive capabilities, have converted them into force multipliers in modern naval warfare. They have become more common in monitoring chokepoints, tracking either surface or subsurface threats, A2/AD mission support, or providing comparable maritime domain awareness across greater horizons. Being very long-range and autonomous, they are operated very often from land bases.

5. Carrier-Launched UAVs

These UAVs are mostly fixed-wing-disposed and include systems such as catapults for their launch and arresting gear for their recovery. They allow further power projection from home waters by an enhancement of ISR and strike capability by a carrier group. Designed for great survivability, endurance, and autonomy, they may provide combat support, early warning, and electronic warfare missions with manned fighter jets. Such drones prove to significantly contribute in preventing a naval task force from going unnoticed while sustaining persistent surveillance and air superiority without jeopardizing the lives of human pilots in possible hostile scenarios.

6. Shipborne UAVs

Very small UAV systems which can be launched from and recovered aboard a surface vessel such as a frigate, corvette, or patrol vessel include shipborne UAVs. These systems are usually launched through such means as pneumatic or rail launchers and recovered by nets, tethers, or automatic landing systems.

For reconnaissance, anti-piracy operations, and maritime interdiction and even environmental monitoring, shipborne UAV systems can be employed. Broadly, they can be thought of as small-scale UAV platforms, which can be handled easily by naval units without major structural modifications.

FIGHTER DRONE LAUNCH AND RECOVERY MECHANISMS AT SEA

For the larger fixed-wing drones, the most widely established type of launch system is the catapult-assisted take-off systems, which are pneumatic, hydraulic, or electromagnetic systems installed on the flight deck of either an aircraft carrier or a large naval ship. In the operation of a catapult, the drone is rapidly accelerated to take-off speed in a very short distance before becoming airborne. These kinds of drones will typically use arrestor systems for landings, which refer to the main hook of a drone catching onto arresting wires on a deck so that it will decelerate and come safely to rest. This mimics the way in which manned aircraft launch and land aboard aircraft carriers; this adoption would suit high-endurance, carrier-launched UAVs like X-47B or proposed naval versions of MQ-25.

For smaller vessels or drones operating in confined environments, VTOL capability offers an alternative that is flexible and compact. However, they operate just like helicopters, which take off and land vertically in space: destroyers, frigates, and patrol ships can all use such drones. These usually are also made small in size in terms of the hardware within them and rely on small rotors or tilt-rotor mechanisms in conducting vertical landing and take-off as helicopters do. Their timestamp would then suffice for immediate use and pick-up, even during bad rough seas and narrow deck area constraints.

Figure 3. Catapult Launch of a Fixed-Wing UAV from a Naval Vessel

Figure 3 shows a fixed wing UAV being launched from the deck of a naval vessel through a pneumatic catapult or mechanical catapult system. Such systems are used for medium to large drones that cannot be launched vertically and need to be assisted to reach the flight speed while being in a restricted deck environment. Catapult launch allows UAVs to be immediately fitted, without needing a runway, into service for sea-based operations in blue water or littoral zones. This is very beneficial for UAVs built to forage or fight over long ranges, and empowers the navy in distributed maritime operations. This figure depicts a conventional fixed-wing UAV being launched from the deck of a naval vessel by a pneumatic or mechanical catapult system. Such systems are most often deployed for renters of medium- to large-sized drones that require assistance in taking off because they cannot do a vertical take-off and cannot attain flight speed in a constraining deck environment. This supports UAVs intended for long-range missions in surveillance or reconnaissance missions, and strikes, enhancing naval capability in distributed maritime operations.

An emerging long-endurance, at-sea solution for these unmanned aerial vehicle operations includes creating drone-carrier ships and associated portable recovery systems. These specially designed vessels will include facilities for UAV operations, flying decks, launch rails, automated retrieval systems, and UAV control centers. Medium-size ships- or even unmanned surface vessels (USVs)-could use a net recovery, skyhook, or precision landing pads employing optical or radio beacons to guide the drones to safety. Thus, these systems are portable enough to allow for the launching of drones on a wider range of naval platforms and the key cornerstones of distributed maritime operations, where UAV support is needed considerably far away from the standard runways or carriers.

OPERATIONAL ROLES AND USE CASES OF FIGHTER DRONES IN MARITIME APPLICATIONS

Fighter drones or military-grade Unmanned Aerial Vehicles (UAVs) have dramatically expanded the operational scope of contemporary naval forces. Their versatility, quick-deployment feature, and compatibility with sophisticated sensors and communication infrastructure make them essential tools for both military operations and support functions. Aboard the sea, fighter drones are not mostly limited to recon and more widen the scope of military with civilian activities alike (Gupta, Bhatnagar, & Mehta, 2021). In the following sections, the major operational applications will be addressed concerning fighter drones above vast water surfaces.

1. Intelligence, Surveillance, and Reconnaissance (ISR)

With Intelligence, Surveillance, and Reconnaissance (ISR) very firmly within naval operations, land and sea-based fighter drones serve primarily in these services. EO/IR-camera-equipped drones provide real-time observation across vast expanses of ocean, while radar and AIS (Automatic Identification System) receiver-equipped drones intercept signals from commercial ships that might be using AIS transponders. ISR is crucial for the detection of any foreign naval movement, the monitoring of commercial or suspect vessels, and the creation of Maritime Domain Awareness (MDA). As opposed to piloted patrol aircraft, UAVs can hang around over points of interest for long hours, even in hostile areas. For example, tactical drones from destroyers or aircraft carriers can monitor enemy ship traffic on their own and send encrypted data to the command center. UAVs also help in intelligence fusion by providing real-time imagery into centralized systems where AI applications analyze vessel patterns, routes, and anomalies.

2. Search and Rescue (SAR)

A SAR operation is one that takes place over the ocean, presenting large and mostly featureless areas in addition to dismal and unpredictable weather conditions. Very promising as Search and Rescue deployments have been fighter drones, particularly those fitted with thermal and multispectral imaging sensors.

According to Wang et al., (2023) UAVs mounted with CNN-based detection software can locate human bodies in sea environments with a great deal of spatial precision-hence making it much more efficient to quickly send drones that can be dispatched to a location in distress, scout large search grids, and guide rescue boats to survivors than the conventional methods. Similar deep-learning approaches have been validated in different sea-rescue scenarios (Ramesh & Jeyan, 2020). Most of the operations are most beneficial for time-sensitive situations of shipwreck victims, persons overboard, or refugee boat sinkings. Some UAVs can even throw flotation gear or supply emergency rations.

3. Maritime Patrolling and Interdiction

Unmanned aerial vehicles, more commonly known as UAVs, have begun taking on the responsibilities of patrolling and blocking certain areas external to international shores, more so, within the exclusive economic zones, and the sea lanes through which vessels pass. They are made in such a way to entail seeking out and reporting illegal activities such as piracy, illegal fishing, arms smuggling, and smuggling. In extreme cases, armed drones can give warning shots or take out threats autonomously or semi-independently with human oversight. As with the land and airspace, UAVs could be scripted to fly on pre-programed routes, and they could also use adaptive algorithms that would enable them to modify their course when anomalies related to maritime traffic detection or alerts from satellites are found (Duan & Zhang, 2014).

4. Anti-Submarine Warfare (ASW) and Strike Missions

Once strictly under the purview of manned aircraft and naval helicopters, Anti-Submarine Warfare (ASW) is coming to include UAVs as auxiliary and complementary assets. UAVs equipped with magnetic anomaly detectors (MAD), sonobuoy-dropping devices, and passive acoustic sensors are under testing to aid in submarine detection and localization missions. In strike operations, UAVs, especially the weapons-capable variants like the MQ-9B Sea Guardian ((Mukherjee, 2018), can target enemy ships, radar installations, and attack craft. Guided munitions or loitering munitions fired from drones afford adaptable, rapid-reaction strike capability at less cost and risk than manned strikes. Swarm tactics are also being discussed in naval strike doctrine, whereby multiple drone units engage or saturate ship defenses, as explored by Leman & Nevalainen, 2025.

5. Environmental and Pollution Monitoring

While their presence in offense and defense, fighter drones also provide environmental safeguarding from pollution. Examples of employing drones in the field of environmental protection include sulfur emissions from ship exhausts, spill detection, tracking the ocean plastic flows, and evaluation of damage to ocean ecosystems. UAVs transported gas sensors, spectral imaging technology, and chemical detection payloads may fly over ship funnels to ensure adherence with MARPOL Annex VI emission standards, while also detecting illegal bilge water discharge or ballasting off coastlines. They are also an aid in environmental effect studies through mapping coral reefs, coastal erosion, and sea habitats from an aerial view, frequently assisting both civilian environmental commissions as well as naval ecological operations (Frederiksen & Knudsen, 2018).

To cut a long story short, the evolution and diversification of combat drones have resulted in their performing a wide range of maritime operations besides the military ones apart from their initial ISR mission. Their value to fight future naval power is indisputable due to their quickness, adaptability, and capability of performing risky tasks while being at a distance in scenarios such as combat search and rescue or, incidentally, environmental stewardship.

COMBAT INTEGRATION AND NAVAL DOCTRINE

Extending the drone missions to flying reconnaissance missions was the first step, and now they have become essential platforms in naval warfare, mounting sensor loads capable not just of precision attack but also autonomous decision-making in real time, streaming data across networks.

The radical shift in operational architecture of naval forces is evident in the full fusion of fighter drones with combat systems. This change is also reflected in the significant evolution of doctrines that are the direct consequence of this transformation. In other words, they will be an integral part of the maritime battle theatres, contributing to the lethality and survivability of naval forces, thus, a complete rethinking of command and control, rules of engagement, and force structure is necessary.

Tactical UAVs in Naval Battlespace Management

Modern navies are increasingly adopting tactical unmanned aerial vehicles to provide real-time situational awareness, assist dynamic targeting and battlefield coordination, and enable multi-domain interoperability.

The sensor coverage of naval vessels has now been greatly extended beyond organic capabilities and correspondingly decreased the response times for emerging threats. Such UAVs are considered vital asset components of modern maritime domain awareness (MDA), which provides constant observation on surface, subsurface, and aerial threats across wide, often-contested maritime zones.

For example, the U.S. Navy has adopted a densely layered UAV integration strategy, including short-range ship-launched drones for localized surveillance, tactical drones used to provide ISR cover in the mid-range, and long-endurance assets capable of coordination with satellites and underwater vehicles. It introduces the concept of layered integration and tactical UAVs as integrated-collaboration components of naval operations, rather than isolated units (Singh, 2016).

Figure 4. Alpha 900 tactical UAV (https://alphaunmannedsystems.com/products/alpha-900/)

Tactical rotary-wing UAVs, shown in Figure 4, have been designed for maritime environments. They serve a critical role in naval battlespace management, providing persistent surveillance, dynamic targeting, and real-time situational awareness. These UAVs feature VTOL capabilities and are perfect for deployment from ships having limited deck space. They prolong the sensor range of naval vessels while being on the ground, air and sea command, they provide the necessary support for the execution of operations in the heavily defended littoral areas.

1. Swarm Warfare and Shipboard Laser Countermeasures

With the rising threat posed by drone swarms—especially heterogeneous swarms combining surveillance, decoy, and strike capabilities—the navies need to come up with new doctrines on how to identify and mitigate threats. Leman & Nevalainen, (2025) discussed the defense efficiency of laser weapon systems (LWS) mounted aboard ships against attacks comprising multi-drone attacks and showed that conventional kinetic defenses might fail against highly coordinated adaptive formations of drones. His research shows that several factors such as density of the swarm, shape of formation, and timing when vectoring the attack play important roles in interception capabilities: hence, intelligent counter-swarm systems driven by AI and real-time feedback loops are needed (Moshiri, Azadi, & Malek, 2020).

Swarm warfare also opens up new avenues for the offense. Fighter drones in a coordinated formation could be combined for distributed strike missions, execute saturation attacks against A2/AD (Anti-Access/Area Denial) zones, or conduct synchronized operations in electronic warfare.

2. Real World Applications by India, Israel, and the United States

Many countries now have active reservations in the operational domain with fighter drone systems, indicating the benefits of UAVs in naval operations. India has similarly developed homegrown UAV programs through initiatives such as DRDO's Rustom and Ghatak platforms, with increasing focus now laid on the integration of UAVs for coastal surveillance and carrier-based reconnaissance. There is still work to be done in power management, sensor miniaturization, and creating secure communication networks, but a significant amount of progress has already been made in these areas, as acknowledged by Nair, (2023).

Israel remains the nation farthest ahead in both autonomous ISR and maritime interdiction systems in operation for UAV development. There is overwhelming operational maturity of Israeli naval drones, including the integration of AI for autonomous target tracking and dynamic threat assessment operations (Nair, 2023).

America has been a pioneer in deploying UAVs across all maritime domains, from MQ-8C Fire Scouts for vertical take-off surveillance to MQ-25 Stingray for carrier-based refueling missions, to Sea Hunter USV, which coordinates UAV–USV missions. This clearly demonstrates a deliberate move toward distributed lethality and unmanned–manned teaming in U.S. naval doctrine.

3. Drone-USV/Ship Interoperability

Interoperability among fighter drones, surface vessels, and unmanned surface vehicles (USVs) is a central selling point for this future naval fleet. Network-centric combatization prescribes all platforms (manned or unmanned) to operate as data nodes within a collective combat cloud. Because an UAV can function either as manned action or as self-directed cooperation with ships and USVs against ISR, targeting, and kinetic or non-kinetic action, it becomes the force multiplier. Examples include real-time hand-off of data by a drone to shipboard battle systems; coordinated loiter patterns over designated sectors; and mission re-tasking depending on sensor input from the surrounding unmanned platforms.

New advances featured in Leman & Nevalainen, (2025) among others, show how feasible, even today, operational use of such systems is. These units enable cooperative swarm ISRs, launch and recovery procedures from moving vessels, as well as AI-guided intervention tasks, thus marking great progress in terms of autonomy and interoperability.

ROLE OF AI AND MACHINE LEARNING IN MARITIME DRONE WARFARE

Artificial intelligence and machine learning have been commonly used to reshape the operational potentials of fighter drones for maritime purposes. One excellent AI task in the context of fighter UAVs is the autonomous navigation and dynamic replanning--the drone's movement continuously are at safeguard, taking into account the environmental ability of water from storms to high seas and interference from electromagnetic sources. For instance, the AI-enabled navigation systems keep the drone flying in changing trajectories by avoiding obstacles and responding to immediate threats like incoming missiles, or jamming efforts.

AI, furthermore, is being constantly optimized to provide target recognition and behavior pattern detection. Utilizing deep learning models like CNN, image data from the visual or thermal channel of a drone can be combined to find, identify and categorize targets absolutely key enemy vessels, submarines, or essentially suspicious maritime traffic. AI also looks forward to provide real-time decision support and mission-level autonomy. In the instance that an AI-managed drone on surveillance duty detected an extremely high-value target, it could switch autonomously from surveillance to pursue or attack mode, share its observations with nearby drones or ships or even call in reinforcements all depending on the previously agreed-upon threat thresholds.

Figure 5. AI-based object detection pipeline from UAV sensors (Gans & Rogers, (2021))

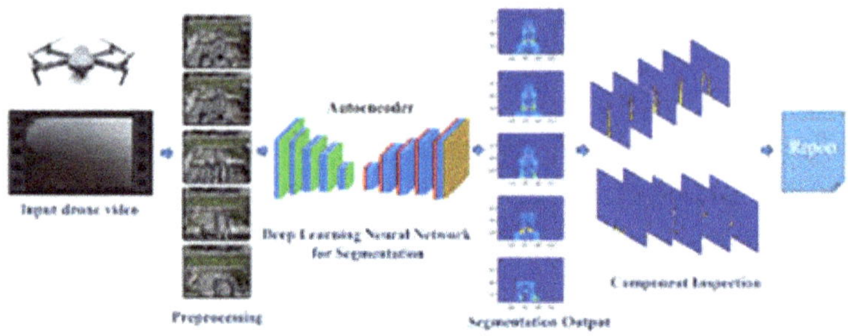

Figure 5 represents the application of deep learning and neural networks in the real-time analysis of drone video footage. The UAV collects a preconditioned type of imagery and feeds it into an autoencoder-based segmentation model. With the aid of this automated action, the act of segmentation recognizes and isolates the object of interest (that is, vessels and infrastructures) for component inspection, and generates autonomous reports. With such functional ability, the awareness of the situation and the operator's exhaustion can be lowered, however, the acceptance still relies on the operator's preparedness and training if there are no appropriate human–machine interaction models, drones will fail in actual operations.

Though a lot has been achieved in the field of car autonomy, the human factor is still very important. The success of on-the-ground drones is mostly due to the operator's training, cognitive readiness, and the trust they have in the AI systems. The Human-Machine Teaming (HMT) model is among those paradigms that point to the idea of human operators who, through augmented reality displays, voice, or gesture-based control as an intuitive interface, manage several drones simultaneously. In fact, a proper training program should involve not only the development of technical skills but also raise the ethical and legal awareness of the users so that they become responsible operators of drones who do not harm the environment. If there is organizational preparedness and well-trained operators, then the failure of advanced maritime UAV systems in their real-world missions is highly unlikely.

This way, AI and machine learning programs will be the foundation of naval warfare through drones. They will gain autonomy, precision, adaptability, and complete fighter drone performance with low human interference and lots of variation in the engagement of complex contested maritime environments.

ROLE OF DRONES IN GREY ZONE MARITIME WARFARE

Grey zone warfare refers to a range of activities that are situated between traditional peace and open armed conflict, these usually are coercive tactics, strategic ambiguity, and acts of non-attribution. In the case of the ocean, unmanned aerial vehicles (UAVs) have become exceptionally efficient tools for carrying out such operations without escalating to a full-scale war. Their ability to stay for a long time, to deny responsibility, and their accuracy are some of the features that make them the perfect means for the projection of power and the undermining of the enemy's influence under the threshold of a conventional military engagement.

Figure 6. AI-Enabled Maritime Surveillance by UAVs (https://sailorspeaks.com/2025/05/29/will-drones-redefine-indias-defense-of-its-seas/)

Figure 6 depicts an advanced drone using radar scanning and an Artificial Intelligence powered automatic identification system (AIS) for real-time maritime surveillance. Grey zone maritime warfare UAVs can indefinitely patrol the waters under dispute and keep an eye on the movement of the vessels (Wang et al., 2023) at the same time, they can also provide the classification of ships and the detection of anomalies without increasing the tensions. By means

of AI, the drone is also able to sift through enormous amounts of data, and then it can give on-the-ground reports, initiatory warnings, and signals for strategy, which are some of the most essential features of grey zone operations. (Klein, 2021).

1. Persistent Surveillance without Escalation:

Drones enable governance bodies to keep a near-perpetual watch over disputed maritime floaters— might be contented Exclusive Economic Zones (EEZs) or territorial waters levying off human patrolled ships or aircraft. So strategic observation and data collecting is backed up with risk minimization of human casualties and escalation. Being unendingly silent, they form part of typical grey zone tactics, thus employing war threshold below pressure methods.

2. Ambiguity and Plausible Deniability:

They exhibit the largest range of ambiguity in their operations due to unmanned drones. In case a drone is downed or otherwise involved in a confrontation, it leads to speculation. At the same time, actors disavow purpose or ownership. That particular use flexibility of the drone is very much in line with an ambiguous attribution grey zone strategy, thus minimizing the chances for retaliatory actions. Drones, which are also probe drones, examine the reaction of the target or collect the intelligence under some plausible deniability cover.

3. Psychological and Strategic Signalling:

Such drone operations, most likely without armament, can conduct psychological operations as well as strategic communication. Just being there over the contested waters means the capability of seeing through the enemy's tricks, and the dominance of the situation-without a real military option. One way for the state to show the world that it will defend its territory and that it adheres to the law of the sea, is to let a drone fly in the zone of dispute without a showdown.

4. Support for Non-Military Maritime Actors:

In this regard, the employment of drone UAVs generally constitutes cooperation with the coast guard, maritime militias, and civilian vessels, which are involved in grey zone illicit activities such as illegal fishing, resource exploitation, and vessel harassing activities. In this fashion, the UAVs provide these actors with real-time intelligence, surveillance, and navigation support, which assists in consolidating control over the maritime domain and dominating through the use of unconventional methods.

5. Counter-Grey Zone Capabilities and Early Warning:

If we consider the other side of the coin, drones may be instruments of grey zone operations; however, they would be very important means for countering the grey zone as well. These UAVs will improve Maritime Domain Awareness through the detection of suspicious activities, intrusions, and irregular maritime traffic situations. Early warning technologies will lead to naval forces being ready for escalated scenarios, stopping illegal acts, and recording breaches for future diplomatic and legal venues.

ETHICAL AND STRATEGIC IMPLICATIONS

The specific use of fighter drones in a naval operation result in a significant advantage while also raising ethical and strategic issues that are quite complex. The system is beyond the main naval and political systems and is not bound by human limitations; hence it can function there. In majority of the cases, they are found in areas with ambiguous international laws and in the zones of the conflict. As they are autonomous, can be denied and have a long-lasting nature, they invent their unique risks of the escalation, accountability, and stability at sea. Here we address the key ethical and strategic issues emanating from the use of maritime UAVs mainly focusing on those which may substantially change the dynamics of escalation in the areas under dispute: the ethical issues regarding autonomous weapon systems and the effects on strategic deterrence and naval stability in general.

1. Drones and Escalation Dynamics in Maritime Zones

The range within which fighter drones can be utilized is naval zones found to produce new escalation dynamics that arc considcrably diffcrcnt from thosc that humanity is used to in manned operations. UAVs are politically and militarily presented as risk-free assets, and consequently, they are either used in a reckless manner or simply as fast and quite normal boats for surveillance, reconnaissance, and maybe even for strikes. Nonetheless, these actions may actually considerably increase the chance of conflict escalation that is unintentional. The incident with the UAV could have been an infringement of the EEZ of another country or an obvious reconnaissance over the maritime territory that is in dispute.

Such behavior might easily be termed aggression—especially if the drones are armed or even autonomous—but where uncertainty about intentions and attribution is involved, all the more so in the case of UAVs, then this will give rise to divergent responses in either party. The situation thus provides avenues in which either party may retaliate in short order, sometimes before the issue even reaches avenues for diplomatic accommodation. The extremely contested areas of the South China Sea or the Gulf are therefore very much prime candidates.

2. Autonomous Weapons and Ethical Use of Force

The deployment of AI technology into a maritime UAV strongly weighs the ethical demand upon autonomous weapon systems. Unlike remotely controlled drones, autonomous UAVs can select and attack targets with little to no human oversight. The ethical questions about that include issues of responsibility, legality, and proportionality for applying force. Under international humanitarian law (IHL), attacks should differentiate between civilians and soldiers and proportionally apply force and accountability (Konert & Balcerzak, 2021). All decisions easy to argue cannot be coded into a machine (Kraska, 2010).

This ethical programming gets even messier in dynamic and sometimes unpredictable maritime scenarios for autonomous decision-making. Also, there is the moral question of logic-based decisions on life-and-death matters when there is little danger for the personnel onboard and severe implications for the other party. Technological advancements that are moving at a very high speed, for example, drone autonomy capability, are almost out of sync with respect to the ethical frameworks, operational constraints, and international treaties that will regulate the use of weapons in a human-compliant way in the context of a humanitarian normative.

3. Strategic Stability and Deterrence Implications

Fighter drones thus interfere with conventional deterrence and strategic stability modeling associated with naval warfare. The very capability of persistent surveillance and precise strike reinforce deterrence along the lines of demonstrating presence and technological preeminence. They, however, seem to picture strategic instability by engaging low-cost escalation and making pre-emption appealing and novel escalation risks (Klein, 2021) on the one hand. An association with little recognition and denial vis-à-vis UAV operations may obscure the line of attribution and obscure an adversary's risk-calculation and response. A drone operator's reluctant nature in mischief may give rise to misinterpretations and trigger overreactions if in a disputed area. Also, by spreading this technology to non-state actors and regional navies, asymmetrical conflicts and kinetic regions of geography are rendered ever more unstable.

CHALLENGES AND LIMITATIONS

Although the fighter drones have evolved greatly and are being rapidly inducted in operations at sea, there are still many technological, operational, and development-type limitations that hinder effective operations of these assets.

1. Power and Endurance Problems

Constraints on endurance are one of the most crucial constraints of UAVs, especially for long-distance missions at sea. Fighter UAVs are sometimes operated over extensive oceanographic areas, which are often far from refueling bases or recovery locations. UAVs(Arnold et al., 2025) powered by batteries, could be smaller and operate quieter, but in terms of energy density and the period of time they can fly, it's below acceptable levels.

Few exist now powered by fuel cells, or using solar power, but they have not been designed to meet maritime presence requirements for ISRs and strikes. With sensor operations and energy requirements, the power reserve is already under strain.

2. Communication Security in GNSS-denied Environments

Most of maritime conditions, especially in contested or hostile zones, are unable to receive GNSS signals due to jamming, spoofing, interference, or cyber means. Any attempt to operate fighter drones in these regions may be hindered due to the inability of loss of signals, data corruption, or global takeover. Though secured alternatives based on inertial navigation systems (INS), visual odometry, and AI path planning might provide some solution, it remains the holy grail to design low-latency and confidential communications over long distances without the need for satellite support.

3. Sea State and Weather Challenges

The maritime environment is typically unpredictable and very often hostile. Adversities to drone stability, performance, and longevity include winds, humidity, saltwater corrosion, rain, fog, and a highly disturbed sea state. (Ramesh & Jeyan, 2020) Fighter drones must withstand saltwater and high-pressure spray exposure and must be stable and under control in gusty high winds typical of high seas. Under rough sea state, take-off and landing become quite difficult, especially for moving platforms like ships, which narrows the operational window and calls for advanced stabilization technology.

4. Skill Set and Domestic Development Hurdles

Skilled human resources are generally required across aeronautics and avionics, software engineering, artificial intelligence, and materials science, but all of these are often in short supply. For example, India where lack of advanced domestic components, heavy dependence on imports for sensors and communication modules, thin access into the global drone ecosystem slowed down self-reliance in UAV development. Poor R&D infrastructure, incoherent policies, and budget constraints are bottlenecks to building a competitive domestic drone industry. Drone deployment case studies indicate that the end result is not always positive. As an example, the first tests of aerial vehicles on ships in rough seas led to multiple failures during the process of getting the drones back, which subsequently caused a delay in the operational rollout. Furthermore, the difficulties in obtaining BVLOS permits have been the cause of slowdowns in the deployment of long-range drones in some naval fleets, even if the drones are technically ready.

CYBERSECURITY AND DATA GOVERNANCE IN MARITIME DRONE OPERATIONS

Fighter drones produce confidential maritime intelligence and send it, which is why they might be exposed to hacking, jamming, spoofing, and interception of data in waters under dispute. Cybersecurity frameworks stress the use of end-to-end encryption, very strong command-and-control channels, and intrusion detection systems. Advanced countermeasures, for example, frequency-hopping spread spectrum, blockchain-based data provenance, and AI-driven anomaly detection, are more and more being brought into experiments with the goal of protecting communications of maritime drones. (Nomikos et al., 2022) Defenses against jamming and spoofing are built on hybrid navigation (INS, visual odometry, and quantum sensing) which is able to provide reliability even in a situation where GNSS is completely denied. Data governance that works well also comprises of data access control, data retention, and interoperability between the fleets of allied nations as part of the policies, which in this way intelligence collected is not only secure but also can be shared without giving away the secrecy of the mission.

REGULATORY TRENDS AND GLOBAL FRAMEWORKS

The authority for drones working at sea is significantly influenced by the changing international and national regulations. The International Maritime Organization

(IMO) has come up with the concept of Maritime Autonomous Surface Ships (MASS) which provides a few basic principles for safety, responsibility, and the compatibility of these ships with the existing laws of the sea. In a similar vein, the BVLOS (Beyond Visual Line of Sight) licensing regimes are an absolute necessity to allow long-range missions of naval drones. The USA, the UK, and India are examples of countries that are testing the frameworks that aim to strike a balance between operational safety and the accountability of the staff. (Kraska, 2010) Through the implementation of these global standards in naval doctrine, drones can be legally and seamlessly integrated into operations, hence the chances of disputes over legality or drones which have been wrongly used in international waters are minimized.

LOGISTICS, MAINTENANCE, AND OPERATIONAL SUSTAINABILITY

Today's naval UAVs are provided with the onboard maintenance systems that assess system health during flight and docking/undoing on carrier ships. These systems then relay real-time alerts about component degradation, overheating, or some anomaly signaling impending failures. Miniaturized UAV servicing stations are now fitted on the carrier decks for quick battery swapping, modular part replacement, and sensor recalibration and refueling for fuel UAVs. In addition, this is done through less strenuous logistics operations for the ship crews and ultimately operational availability is maximized. (Leman & Nevalainen, 2025) Furthermore, the management of spare parts and the implementation of predictive maintenance represent the most important aspects of the drone sustainment system. The drones have some fragile electronic parts and sensors that can be affected by environmental factors; and while they may be damaged by salt spray and moisture, for the most part, they are damaged by mechanical stresses.

With the AI-driven maintenance systems in operation, navies can now predict when certain components are close to failure based on their usage, flight data, and exposure to the environment. Predictive analytics allows for proactive part replacement, thereby minimizing risk in-flight and optimizing maintenance cycle placements. AI systems can also automate onboard spare parts inventory tracking while on board the ship to ensure important components are stocked and resupply chains are managed in advance.

Fighter drones are therefore cost-efficient when compared to manned aircraft. While UAVs have high initial procurement and integration costs, long-term operational costs are usually lower because no training is required, there is no risk to the pilot, and maintenance cycles get simplified. The long-term integration of fighter drone operations into naval personnel, therefore, would be conditioned by logistics, maintenance, and sustainability. Well-established AI-driven diagnostics, modular hardware, and effective support infrastructure permit the maintenance of a persistent drone presence at sea at lowest cost and highest readiness for naval forces. With time, more back-end support systems alongside the drones will come into play in the operational success of the task forces. (Leman & Nevalainen, 2025).

FUTURE DIRECTIONS AND EMERGING TRENDS

Sea fighter drones are expected to lead to further advancements in smart autonomy, endurance, and multi-domain integration and coordination with the use of emerging technologies at a deeper level. These innovations from quantum sensing to swarm logic and hypersonic platforms will transform naval power (Ruiz Estrada, 2024) but they need to be weighed against risk scenarios, regulatory hurdles, and uneven market adoption to present a realistic roadmap. Understanding these emerging trends is very important to setting the future naval strategies and policies correctly. On the one hand, technological innovation has almost infinite possibilities to delve into. However, the decision to take these chances depends mostly on market dynamics and industry readiness. Some (2022-2025) predictions expect that the worldwide maritime UAV market will have a compound annual growth rate (CAGR) of more than 12%. On top of that, among the regions, they recognize Asia-Pacific and North America as the two that will be leading the market purchasing.

From an economic point of view, unmanned systems are a smart choice as is clearly shown in the cost-benefit comparisons which reveal that these solutions provide 30-40% less long-term operational costs than crewed patrol aircraft. However, the integration of the upfront still remains quite a high one. The private sector which includes shipbuilders and defense contractors is not only getting more involved but also speeding up the process and making it easier for the development of such new concepts as modular drone-carrier and hybrid UAV–USV platforms. Moreover, this fact is raising the role of commercialization trends as a key for future adoption even further.

1. Integration of 6G Networks with Edge Computing:

Fighter drones will benefit immensely from extreme speed communication with extremely short time delays that will come from the new 6G systems. These drones will link up anywhere on earth with ships, satellites, or command centers in the farthest regions of an ocean. Decisions of these drones will be based on processing data from onboard sensors without needing external servers, thanks to edge computing, which reduces reaction time and leads to more independent operations in denied GPS environments or hostile.

2. Autonomous Mission Planning & Self-Healing Swarms:

Future fighter drones will incorporate artificial intelligence into the planning and real-time adaptation of missions. Instead of executing a pre-set mission by a human operator, the drones will adapt and react to changes in conditions surrounding the mission, e.g. due to weather change or new threats. When it comes to more self-healing swarms' deficiencies, redefining capabilities after a unit or more have failed might allow the mission to occur continuously, without interruptions resulting from those failures, because of an increase in resilience and operational effectiveness.

3. Underwater UAV-Aerial UAV Coordination

The coordination of aerial drones and autonomous underwater vehicles (AUVs) will traverse combinations and multiple domains in naval activities. For example, on aerial scouts or communication relay functions, AUVs will perform good stealth underwater missions, such as mine detection or tracking submarines.

As a result of such a combined method, the company is more aware of the situation and able to apply the mission in a flexible way. Interactions between different operational domains can be specified in certain instances like maritime security, bottlenecks, and sub-surface warfare (Arnold et al., 2025).

4. Quantum and Hypersonic Drone Technologies:

Quantum navigation and sensing technologies provide extremely high precision of positioning and detection that can't be matched in GPS-denied environments. This means that the drones will be able to navigate through and perceive underwater and low-visibility threats much better. Simultaneously, additional hypersonic drones over Mach 5 speed are thought to be taken into account for rapid-response missions, delivering intelligence or strike capabilities before any conventional system and changing the pace and reach of maritime operations.

5. Integration with Space-Based Assets:

Future fighter drones will increasingly connect with space-based systems such as military satellites or high-altitude pseudo-satellites (HAPS), which will function as global communication relays in supporting long-range, all-weather surveillance and navigation enhancement. This ensures continuous connection to UAVs over oceans, even in regions devoid of local command infrastructure.

6. Human-Machine Teaming (HMT):

HMT will perforate large-scale future drone naval operations. Using intuitive control interfaces enabled via augmented reality or voice commands, pilots or naval operators will simultaneously supervise a number of drones. This partial automation of control would relieve the operational burden placed on the human operator, hasten decision-making, and grant the drone autonomy over high-level tasks with little human input required.

7. Enhanced Energy Systems:

Most energy systems of the future will address extended endurance, which still poses a core challenge. Fighter drones may soon adopt hydrogen fuel cells, solar-assisted charging, or of wireless power transmission technologies. While these developments will improve the endurance in flight, cut down the need for refueling at sea, and allow for mission availability that supports persistent surveillance or strike roles, they will not remove some operational missions' dependency on being created (Nair, 2023).

8. Modular Payload Systems:

Modular design will allow interchangeability of mission-specific payloads such as sensors, munitions, or communication relays for the drones. This re-configurability is supposed to reduce the number of different drone types required and afford a single platform to support multiple roles, thus meaning reduced turnaround times and enhanced flexibility in deployment from ship-based platforms for naval operations.

9. AI-Driven Electronic Warfare Capabilities:

Future drones will most probably come equipped with AI-enabled electronic warfare capabilities to detect, jam-up, or spoof the enemy radars and communications. Such capabilities will become precious for operations in contested maritime zones in which possibly ordinary electronic warfare platforms could become vulnerable. Fighter drones will, thus, materialize as more offensive and defensive mobile electronic warfare platforms, allowing for both the suppression of enemy defenses while keeping crewed aircraft out of harm's way and allowing killing missions to enemy bases from long distance.

10. Underwater Launch and Recovery Systems:

Under development are systems that can launch an UAV from a submarine or from an underwater platform. In this way, secret drone launching can be done during a hidden surveillance or a sudden attack. By adding this element to the operations at sea, naval stealth strategy has in effect been revolutionized which means that it should be possible for the aerial units to carry out their missions without the risk of revealing the surface assets. Such technology does not only keep the enthusiasm alive for the innovation side of things but also ensures the necessary level of practical readiness, thus constituting a realistic plan for the future of maritime drone integration.

CONCLUSION

Fighter drones have become a paradigm-shifting element in the contemporary naval warfare environment, largely affecting the contemporary navies' strategic as well as tactical levels operations. Moreover, unmanned systems are rapidly expanding their applications in naval operations; seemingly, these could be utilized for surveillance, reconnaissance, strike missions, and environmental monitoring, thus reshaping the paradigm of persistence, precision, and operational reach at the sea. This chapter reviews the multifaceted capabilities of fighter drones, their major impacts, and the potential future of these in maritime security and naval doctrine sectors. Through autonomous or semi-autonomous missions, such as ISR, strike operations, environmental monitoring, and humanitarian aid, fighter drones have demonstrated the capabilities to become indispensable in modern naval warfare. They are also safe and sound for various mission profiles from grey zone engagement and underwater surveillance to swarm-coordinated attacks and strategic deterrence with their dual-use nature, which is still hardly discerned by most. Besides the positives, fighter drones show performance limitations in terms of operational endurance, vulnerability during GNSS-denied missions, weather disruptions, and a shortage of qualified personnel for their development and employment. Arguing for the ethical implications of increased autonomy in weapon systems is, nevertheless, a difficult balancing act that requires putting forward issues of accountability, law compliance, and proportionality in respect of the use of lethal force in operations under contestation. The next era of naval drones will benefit from the rise of novel technologies such as 6G connectivity, quantum navigation, AI-enabled swarms, and underwater-aerial integration, which together will boost intelligence, survivability, and flexibility. While AI-powered maintenance systems, modular components, and agile logistics will be the factors behind the drones' longevity, at the same time, the international naval doctrine has to be modernized together with global policies to regulate the ethical and strategic use of unmanned systems at sea.

REFERENCES

Arnold, A. C., Gasparoto, H. F., Brunet, J. P., Vega, P. J. S., & Violante, P. (2025). Challenges and Innovations in Drone and Autonomous Vehicle Simulators and Controllers. *OCEANS 2025 Brest*, 1-10.

Duan, G. J., & Zhang, P. F. (2014). Research on application of UAV for maritime supervision. *Journal of Shipping and Ocean Engineering*, *4*, 322–326.

Frederiksen, M. H., & Knudsen, M. P. (2018). Drones for offshore and maritime missions: Opportunities and barriers.

Gans, N. R., & Rogers, J. G.III. (2021). Cooperative multirobot systems for military applications. *Current Robotics Reports*, *2*(1), 105–111. DOI: 10.1007/s43154-020-00039-w

Gupta, A., Bhatnagar, A., & Mehta, A. (2021). Application of drones in maritime industry (firefighting). *Bulletin of Marine Science and Technology*, *14*, 59–69.

Klein, N. (2021). Maritime autonomous vehicles and international laws on boat migration: Lessons from the use of drones in the Mediterranean. *Marine Policy*, *127*, 104447. DOI: 10.1016/j.marpol.2021.104447

Konert, A., & Balcerzak, T. (2021). Military autonomous drones (UAVs)-from fantasy to reality. Legal and Ethical implications. *Transportation Research Procedia*, *59*, 292–299. DOI: 10.1016/j.trpro.2021.11.121

Kraska, J. (2010). The law of unmanned naval systems in war and peace. *Journal of Ocean Technology*.

Kumar, S., & Sharma, N. (2022). Emerging military applications of free space optical communication technology: A detailed review. [). IOP Publishing.]. *Journal of Physics: Conference Series*, *2161*(1), 012011. DOI: 10.1088/1742-6596/2161/1/012011

Leman, J., & Nevalainen, P. (2025). The Future of Unmanned Naval Systems: An Overview Of Efficiency Enhancements and Warfare Upgrades for Maritime Drones. *Robotics and Autonomous Systems*.

Mukherjee, T. (2018). *The role of artificial intelligence in naval operations*. ORF Observer Research Foundation.

Nair, C. D. J. N. (Ed.). (2023). *Emerging Defence, Maritime and Aerospace Technologies by DRaS*. Highlyy Publishing LLP.

Nomikos, N., Gkonis, P. K., Bithas, P. S., & Trakadas, P. (2022). A survey on UAV-aided maritime communications: Deployment considerations, applications, and future challenges. *IEEE Open Journal of the Communications Society*, *4*, 56–78. DOI: 10.1109/OJCOMS.2022.3225590

Ramesh, P. S., & Jeyan, J. M. L. (2020, December). Comparative analysis of the impact of operating parameters on military and civil applications of mini unmanned aerial vehicle (UAV). In. AIP conference proceedings: Vol. 2311. *No. 1* (p. 030034). AIP Publishing LLC. DOI: 10.1063/5.0033989

Ruiz Estrada, M. A. (2024). A New Advanced Military Drones Systems: Prototypes Modeling. *Available at SSRN* 4912793. DOI: 10.2139/ssrn.4912793

Singh, A. (2016). Unmanned and autonomous vehicles and future maritime operations in littoral asia. *ORF Special Report, 28.*

Taylor, A. B. (2021). *Counter-Unmanned Aerial Vehicles Study: Shipboard Laser Weapon System Engagement Strategies for Countering Drone Swarm Threats in The Maritime Environment* (Doctoral dissertation, Monterey, CA; Naval Postgraduate School).

Wang, J., Zhou, K., Xing, W., Li, H., & Yang, Z. (2023). Applications, evolutions, and challenges of drones in maritime transport. *Journal of Marine Science and Engineering*, *11*(11), 2056. DOI: 10.3390/jmse11112056

Chapter 7
Smart Intrusion Detection System for Restricted Maritime Zones:
Using Motion Analysis and Spatio-Temporal Processing

C. V. Suresh Babu
https://orcid.org/0000-0002-8474-2882
Hindustan Institute of Technology and Science, India

E. Thirumalai
Hindustan Institute of Technology and Science, India

S. Ragunath
https://orcid.org/0009-0002-8028-907X
Hindustan Institute of Technology and Science, India

S. Sujin
Hindustan Institute of Technology and Science, India

ABSTRACT

This chapter per the authors presents the design and evaluation of a smart intrusion detection system tailored for restricted maritime zones, where safeguarding assets against unauthorized access is a pressing challenge. The objective is to integrate motion analysis with spatio-temporal processing to enhance surveillance reliability in dynamic marine environments. The proposed framework employs background subtraction, optical flow techniques, and deep learning models such as CNNs and LSTMs to capture movement patterns over time. A prototype was developed and

DOI: 10.4018/979-8-3373-3546-9.ch007

tested using drone-captured video feeds, demonstrating improved accuracy, reduced false alarms, and efficient detection under varying conditions. Results highlight the system's potential to strengthen maritime security operations by providing real-time situational awareness and supporting regulatory compliance. The chapter concludes that such AI-driven surveillance solutions are essential for advancing autonomous maritime security, with implications for ports, naval facilities, and offshore platforms.

1. INTRODUCTION

1.1. Overview of Maritime Security Challenges

The fluid nature of maritime security is a result of increasing threats; piracy has now turned into a business, apart from illegal fishing, smuggling, and various illegal activities that pose risks to international sea routes, warships, and human life close to offshore platforms for energy. Traditional monitoring systems, namely coastal radar systems, and satellite monitoring systems, are usually inadequate, due to coverage deficiencies, a high rate of false alarms, and incompetence to track time-variable sea conditions (Venskus et al., 2019). As vessels increasingly become autonomous and the nature of threats continuously changes, it becomes a vital necessity to implement smart systems that would ensure end-to-end real-time situational awareness. The natural phenomena and the earth-based variables that limit the vision based maritime surveillance systems would be waves, fog and changing light. The complexity rises at the coasts and crowded places where image clutter and vessel concentration reduce the levels of reliability in detection. These constraints pose the need of AI-driven surveillance that uses spatio-temporal modelling and machine learning to recognize intrusions in valuable maritime areas (Bloisi et al., 2016).

1.2. Importance of Restricted Zone Protection

The maritime zones are classified as restricted in order to support national interests that are vital and they include; naval activities, energy infrastructure, biodiversity in the marine and economic rights. Such areas include naval installations, offshore oil platforms, underwater cables, and marine sanctuaries; unauthorized intrusion to any of these locations can instigate serious geopolitical, ecological or economic effects (Talpur et al., 2025). These areas are sensitive and hence the normal perimeter-based monitoring methods cannot guarantee real time reliable intrusion detection.

Unauthorized vessels and underwater vehicle operations or suspicious loitering may go unnoticed since the incident can be out of radar coverage and the observation ability of human watch personalities is limited. The resulting combination of increasing dependence on, as well as increasing use of, automated, intelligent surveillance systems that are continuously adaptively and autonomously operating for security purposes creates new challenges. The detection using AI, especially by analyzing the motion in time and space, is a scalable solution that can observe and interpret movements in large oceanic expanses and enables the detection of a security breach early (Ahmed et al., 2022)

Defending of restricted areas is both a military necessity and an international maritime requirements law of protecting the international laws that govern the oceans and seas, the UNCLOS and the IMO directives. Utilizing AI in monitoring of such areas thus enhances the national resilience and complies with global conventions of maritime safety and security.

1.3. Rise of Smart Surveillance and AI-Based Detection

Sea regions host more issues such as piracy, illegal ship operations, smuggling, and illegal entry to offshore platforms. Traditional monitoring systems such as radar, sonar, and satellite surveillance do not usually suffice in today's maritime security requirements. Primary concerns regarding such systems include their low coverage, high rate of false alarm, and failure to adapt to varying sea states (Venskus et al., 2019). These issues are particularly significant in restricted sea regions such as where illegal operations may present significant geopolitical, economical, and ecological implications (Talpur et al., 2025).

Latest spatio-temporal analytics and artificial intelligence technologies have improved considerably real-time analysis, problem detection, and background noise mitigation (Zhang et al., 2024). Of course, these technologies pose beyond performance technical challenges too. High cost of hardware, power for edge equipment, and system longevity in harsh sea environments present immense barriers to adoption (Nikolic et al., 2023). Non-technical issues like data governance, compliance with global regulations (like GDPR), liability for losses in case of failure, and ethical issues of autonomous monitoring remain inadequately considered in the literature (Globisch et al., 2024). This chapter shall address the technical and contextual challenges by proposing a Smart Intrusion Detection System (SIDS) that integrates motion analysis, spatio-temporal modeling along with deep learning in an accountable and environmentally friendly framework. By setting out such a system in the dual frames of technological feasibility and regulative-ethical responsibility, such an approach proposes a comprehensive solution to restricted maritime area protection.

1.4. Scope and Purpose of the Chapter

The chapter contributes to the discussion of high technology security and the spread of AI-driven surveillance mechanisms by formulating an all-rounded description of the design, deployment, testing, and analysis of a Smart Intrusion Detection System (SIDS) that is specifically envisaged to operate in restricted maritime areas. The aim is both to fill the gap between motion-detection theory and practice in dynamic, high-risk marine environments and to illustrate how spatio-temporal motion analysis, optical flow methods, and deep learning architectures, including convolutional neural networks (CNNs) and long short-term memory networks (LSTMs), may be integrated to real-time detection of unauthorised activity in videos captured by a drone (Liang et al., 2024).

The resulting system can be deployed on edge devices (e.g., NVIDIA Jetson Nano) and, therefore, can enable offshore and remote processing with low-latency and high-accuracy. In addition to the new prototype architecture, the study provides a quantitative performance analysis, whereby the results of the system were compared to traditional surveillance strategies.

The report also focuses on the practice of practical application, describing the training of models, the interaction of the system, and the creation of alerts, and emphasises strategic applications, requiring protection of the port, naval base, and offshore rigs. As a result, the chapter can serve as a contribution to research as well as a basis of future maritime AI innovations. Our conceptual framework—combining optical-flow motion analytic with spatio-temporal modelling—directly indicates the practical case studies and prototype projects discussed in the subsequent chapters. By embedding deep-learning architectures in practical case studies, from drone-sourcing monitoring of ports to on-or-offshore physical or cyber security, we demonstrate theoretical improvements applied to real-world scenarios clearly showing an outcome for impact (Suresh Babu, C. V., Andrew Simon, P., et. al. 2023). This connection between our conceptual framework and the empirical research evidence: satisfies the aims of the chapter as a contribution to maritime innovation and demonstrates the review theme of the edited book about providing actionable AI-lead security proposals.

2. BACKGROUND AND MOTIVATION

2.1. Evolution of Maritime Surveillance Systems

In the past several decades, maritime surveillance has made significant progress, shifting away to use of conventional radar and sonar systems to AI-based multi-sensor fusion systems. The domain awareness was initially based on the human eye and line of sight radar and although it was effective in detecting big vessels, it was not effective on small ones or non-cooperative targets and exerted its limitations in congested or unfavourable environmental conditions (Bloisi et al., 2016). These shortcomings created a need to adopt Automatic Identification Systems (AIS), satellite images and coastal video surveillance, thus creating hybrid networks that are long-range tracking networks. However, the old systems were more often than not time lagging and limited depending on human tiredness and signal congestion. In the 2010s, automation of target detection, anomaly classifications, behaviour prediction using computer vision and machine learning techniques would allow systems to shift from passive situational awareness to active intelligence generation and maritime surveillance to become a dynamic decision-support capability. Most recently, the introduction of drones, deep learning, and spatio-temporal analytics has paved the way to smart systems that have the ability to identify the motion patterns and produce the threat classification autonomously, and therefore enable the possibility of scaling their intrusion detection to real-time operations in restricted maritime areas (Talpur et al., 2025).

2.2. Challenges in Intrusion Detection in Marine Environments

Intrusion surveillance is complex in the maritime domain. Oceanic ambiguity, sensor inadequacies, and illicit ship paths contribute to the challenge. Unlike cities, the marine environment is fluid and dynamic. Swells, weather, and reflected sunlight interfere with predictable visual tracking and classification. Furthermore, AIS systems suffer from trust and reliability issues. Spoofing or Loss of Signal, from small craft or otherwise, defies the reliability of conventionally used monitoring techniques. This context requires higher-strength systems that do not solely rely on AIS information. There is difficulty due to changes in spatial scale, long ranges, and barriers, and it is challenging to transition from observed motion to detection, tracking, and classification. Complications for these issues result for vision- and rule-based solutions. Delay in computation also complicates many of the smart detection systems deployed on vehicles like drones or coastal sensors. Edge devices are severely limited in their computational capacity, yet expected to produce alerts almost instantaneously. In order to reduce such drawbacks, the recent research employs Kalman filtering, DBSCAN clustering, and deep learning techniques thus improving predicting and anomaly detection even in the noisy marine context.

A significant limitation to moving forward with AI-based maritime intrusion detection is the lack of annotated datasets for restricted areas, like naval bases or offshore platforms. The publicly accessible datasets in the literature (e.g., Singapore Maritime Dataset; Prasad et al., 2017) do not contain any data, for example, about covert loitering and small-craft intrusion. As such, the datasets are limited for model generalization. Synthetic augmentation may assist in bridging this gap, but synthetic augmentation will not account for real life sea states and lighting conditions or multi-angle behaviors of vessels. It is ultimately crucial to have multiple, standardized, multi-modal (e.g. combining video, AIS trajectories, and radar metadata, with expert-annotated intrusion labels) maritime datasets in order to provide robust training and a fair benchmark. These challenges establish the problem context driving the development of advanced motion analysis techniques described next.

2.3. Relevance of Spatio-Temporal Processing

Israel used IoT-enabled sensor networks that include AIS receivers, coastal radar, fixed and UAV-optics cameras and ambient sensors, to perform real-world deployments with unobtrusive, near continuous maritime domain awareness (i.e., IMDA and Mk. II). The edge computing nodes (e.g., NVIDIA Jetson Nano; Shi et al., 2016) provided lightweight spatio-temporal inference pipelines that filter noise and detect objects on the IoT gateway, document motion at the edge and update the classifier on the cloud server; secure MQTT or UDP protocols provided elapsed time from input to output of < 200ms end-to-end latency and preserved battery life on the unmanned platforms. The static frame-by frame object detection within maritime surveillance is often ineffective to identify abnormal behaviour with any precision. Due to that it is critical to use spatio-temporal processing that integrates space-based descriptions (location, size, shape) with time related data (motion trajectories, velocity, persistent patterns) and thus improves the accuracy of the intrusion detection (Suresh Babu, C. V., Suruthi, G., et. al. 2024). Once the system has been trained, it will recognize the baseline maritime behaviours, for example, patrolling or docking and anything beyond this, for instance, unusual loitering or rapid movement in a no-go zone, will be classified as anomalies. Consistent with the types of spatial feature extraction and temporal feature modeling eg Convolutional Neural Networks (CNN's), Long Short-Term Memory (LSTM) networks have been proven to be effective and have outperformed other methodology in trajectory and behavior prediction and classification for many studies. Moreover, the analysis of movement will be separately done with an optical-flow algorithm, using multimedia imagery/non-instructional movement, which would track the motion from pixel to pixel across sequential video frames (Zhang et al., 2024). This approach should provide a substantial enriched level across the domain within the surveillance, but particularly in unstructured maritime domains where anomaly identification should be in real-time. All the technologies referenced will contribute to Maritime domain security in fundamental ways, as the systems will provide added levels to filter-out false positives, and also increased opportunities to respond proactively or escalated to security events.

2.4. Use Cases in Ports, Naval Bases, and Offshore Platforms

Spatio-temporal surveillance systems are establishing themselves as useful in strategic locations in the maritime domain to address threats that come from piracy, illicit trafficking, terrorism, and illegal entry of vessels. Ports as key junctions of trade and supply chains require real-time object tracking and object anomaly detection in order to protect the cargo and avoid unauthorized docking. The video analytics, automatic identification system (AIS), radar fusion, are now used in monitoring congested terminals through the smart systems (Nikolic et al., 2023). In naval ports, where high-security standards are necessary of a military level, AI-based surveillance can detect vehicles that move slowly, or loiter, and exhibit abnormal behaviour. The use of spatio-temporal algorithms allows distinguishing between normal training operations and actual intrusions even when there is poor visibility or poor visibility at night. Oil rigs and renewable energy platforms are offshore structures, which face increased security threats because they are geographically isolated. Unification of drone surveillance with intelligent detection systems provides on-demand surveillance and allows quick actions in response to any suspicious maritime traffic. Collectively, these applications vouch on the need to have smart intrusion systems applicable to multiple contexts and, more importantly, the need to have scalable, real-time, and weather-tolerant detection networks spreading across maritime space.

The Maritime, and Port Authority of Singapore, has managed an AI-enabled CCTV mesh with real-time anomaly detection, which was trained on local vessel traffic and issued unauthorized docking alerts with an accuracy of over 95% and a 50% reduction in false alarms (Zissis & Lekkas, 2017). Within Europe, coastal authorities have conflated footage from drones with AIS and radar data when providing information about anti-piracy patrols and successfully detected unsolicited vessel approaches 85% of the time, and within the 5km monitoring space—cutting the response time to unauthorized approaches by one third (Prasad et al., 2017). Offshore energy operators in the North Sea developed edge-based intrusion networks across several platforms, to process thermal and optical imagery on devices powered by NVIDIA Jetson, to identify small-craft loitering with an above 92% true-positive rate and zero false negatives for six months (Nikolic et al. 2023). These case studies show that AI-enabled maritime surveillance has been implemented on a variety of operating contexts.

3. RELEVANCE TO THE EDITED BOOK THEME

The incorporation of AI-driven environmental monitoring on board ships needs to be governed alongside new ethical and regulatory parameters. The International Maritime Organisation (IMO) has been providing insight into its overarching regulation of the new area of Maritime Autonomous Surface Ships (MASS), including levels of autonomy, in addition to safety issues for both ship and shoreside systems. In another resolution, the IMO now requires cyber risk management as in decision MSC.428(98). In the case of coastal monitoring, owners also must enter the perspectives of non-maritime targets and data governance policies (e.g. GDPR) that regulate geofencing and retention periods. In order to comply with laws, regulations or policies, establishing an effective legal framework with the operational action standards, procedures and rules for lawful, secure and ethical monitoring remains paramount to compliance. Smart intrusion detection systems for confined maritime regions are consistent with and complementary to the present trends in maritime security and autonomy. While formerly universally offering their autonomous vessels and uncrewed autonomous systems (UAS) operations in delivering surveillance services in ports, naval bases and offshore installations; and particularly fiendish systems capable of flawlessly identifying, and intervening in situations in the moment continue to priceless. Given the dynamism of the spatio-temporal analysis of motion, and equivalently artificial intelligence, enhances situational awareness in enabling timely identification of threats in the form of unlawful vessels or suspicious activity. These advancements represent a public good in coastal safety and security, commerce, and the legitimacy of global shipping regimes, while likewise advancing broader counter-piracy efforts. These systems leverage state of the art video analytical tools, sensor fusion, and AI models to meet research and development gaps in the current smart ocean monitoring technologies while creating the next generations of smart ocean monitoring technologies (Globisch et al., 2024).

3.1. Integration with Maritime Drones and Autonomous Vessels

The use of unmanned aerial vehicles (UAVs) and unmanned surface vessels (USVs) has changed the nature of maritime surveillance providing coordinated multi-domain coverage. UAVs can be deployed and travel over very large areas of sea quickly to gather intelligence and USVs can provide persistent near inspection and interception. This collaborative function improves target tracking, improves integration of sensors, and can then inform autonomous decision making process, which is important when considering threats such as piracy, illegal fishing and entering ports incorrectly. A review of the current knowledge of UAV and USVs working together, within an AI-enabled space, indicates that this is important in generating greater detection capabilities in maritime security operations, and can therefore improve the detection and response times (Yao et al., 2019).

3.2. Alignment with Technological Innovations in the Maritime Sector

We are experiencing a period of active transitions, particularly within the maritime sector as a result of, within the maritime sector, autonomous vessels, unmanned aerial vehicles (UAVs) and AI based decision support. These platforms and associated technologies will facilitate the ongoing monitoring, under surveillance, of real time performance and improved situational awareness in maritime areas of interest through detection of anomalies. The complementary application of spatial-temporal analysis coupled with AI improves the capability to gain evidential case detections of unauthorized intrusion in adverse conditions, whilst edge computing capability also allows decisions to be made expeditiously on-site, without saturating the cloud infrastructure with excess data. The collaborative aspect of these technologies takes maritime surveillance to an equal sophisticated level as broader Industry 4.0 Projects, such predictive analytics, IoT assisted port management, and autonomous navigation safety systems. This type of integrated technology will enhance security capability, improve the efficiency of operations, and reduce the amount of time taken to manage incidents at sea (Balduzzi et al., 2019).

3.3. Strategic Importance for Coastal and Port Authorities

Beaches and maritime stewards are struggling with ever-changing volumes of illegal maritime activities, including drug trafficking, illegal fishing, or potentially terrorism. AI-based intrusion detection systems support situational awareness for environments where people are present and at risk by decreasing reliance on human mode of observation (Suresh Babu, C. V., Suruthi, G., et al. 2024). These systems have been trained to observe for unusual motions in populated urban and rural environments, with high accuracy and speed potential, even under poor visibility, weather, etc. These systems enable actionable situational awareness in real-time to minimize the human burden on decision-making, as they have spatio-temporal processing and motion profiles recognized through learning. From an operational standpoint, the speed of detecting and informing authorities about events for decision-making and immediate action across multiple responding agencies is the greatest benefit of these systems. The real-time situational awareness can be based on drone or fixed cameras with footage, facial frames, etc. that authorities will use, and help authorities recognize problems before operations or other events create security breaches. Additionally, these systems can be paired with Vessel Traffic

Management System, to create a secondary layer of awareness regarding vessels that may be operating illegally. The combination of these technologies and camera systems with the vessel traffic management systems in use creates maritime domain awareness for the national security apparatus and thus values the economic stability (Zissi et al., 2015).

4. RESEARCH QUESTIONS AND PROBLEM DEFINITION

Formulating specific research questions is critical to designing effective smart maritime intrusion detection systems which can accommodate both environmental variability and vessel navigation capabilities, as well as processing limitations. Continuous threats, for example incursions into restricted zones for unauthorized activities such as fishing or smuggling, require in-time processing of vessel tracking data and accurate and timely anomaly detection. The three research questions that are explored in this chapter are: (i) how to deal with the unavoidable uncertainties that come in motion analyses, such that vessel behaviour is represented in a realistic manner; (ii) how to determine which spatio-temporal features are useful to distinguish legitimate from suspicious trajectories; and (iii) how to assess more robustness and in-time detection while more accurately lowering false alerts posed in a variety of maritime situations. The larger challenge in innovation is how to integrate spatio-temporal analytics into particular operational maritime surveillance systems (Mascaro et al., 2014).

4.1. Key Research Questions

When planning an innovative intrusion detection system in the limited maritime zones, there are several key questions that need to be addressed, which will inform system design, algorithm selection, and deployment in the field. To begin with, the researchers need to figure out how to detect and classify intrusions in the conditions of adverse maritime environmental factors, including reduced visibility, harsh seas, and nighttime functions (Riveiro et al., 2008). Another question involved how spatio-temporal analysis would help in improving the performance of detecting the things than the use of traditional frame-by-frame motion detection especially in identification of acceptable activities and suspicious ones. Furthermore, one must explore what can be done to optimise real-time processing of data streams of drones, coastal cameras and AIS systems in terms of reducing latency and error without losing accuracy. Last, the research has to resolve the problem of integrating data between heterogeneous sources of data, such as radar, optical sensors, and infrared cameras to create a single coherent model of situational awareness. The above questions aim at ascertaining that the proposed system meets technical and operation needs of maritime security.

4.2. Statement of the Problem

The protection of maritime restricted areas is faced with the constant issues that arise due to changing nature of security threats including illegal fishing, piracy, smuggling, and incursion of unauthorized vessels. The traditional monitor processes' effective radar and AIS (Automatic Identification System) sources are usually incapable of identifying non-cooperative or spoofed vessels. A number of environmental factors, i.e., fog, sea clutter, visibility, etc., not only hinders the vessels from being accurately detected and tracked, but it also intensifies the ship's latent vulnerability. Poor levels of smart and multi-sensor integration are part of the problem with lethargic at-sea surveillance management tools today and how a situation's awareness and reaction times are insufficient. Each environment is unique, but artificial intelligence strengthened with optical sensing and real time spatio-temporal analysis appears to be a good toolset for early anomaly forecasting and to add more meaningful collection capacities to respond to the threat (Tu et al., 2020).

4.3. Need for Smart, Real-Time Intrusion Detection

As surveillance environments become increasingly sophisticated, it has raised the need to develop intrusion detection systems that deliver real-time utility and information in dynamic environments, while also quickly and accurately addressed threats (Suresh Babu, C.V., Suruthi, G., et. al. 2024). The systems thus integrate the spatio-temporal analytics and concurrent stream processing of video and sensors, and allow decomposing flows and detecting anomalies in movements with a high degree of accuracy. Through real time processing pipelines and machine-learning algorithms, modern platforms can adapt to receive constant input and learn it, thus, minimizing false alarms with no reduction in the speed of reaction of the system. Moreover, the introduction of the optical-flow analysis and background-subtraction methods will result in a rise in situational awareness because subtle or gradually traveling intrusions will be detected earlier (Kong et al., 2005).

5. OBJECTIVES AND SCOPE OF THE CHAPTER

The present chapter outlines a smart intrusion detection and protection structure in restricted maritime zones with the combination of motion analysis and a spatio-temporal precinct to achieve greater security. The framework can be used to detect abnormal vessel behavior and initiate counter measures early by collecting AIS data and fusing it with sensor-based tracking. It allows real-time data acquisition, trajectory prediction, and anomaly categorization, and it will be adapted to different maritime conditions and comply with international safety standards. The method is specifically applicable to coastal monitoring, naval warfare and port side information, where there is a high density traffic, environmental conditions and complicated maneuvering (Zissis & Lekkas, 2012).

5.1. Objectives

The proposed smart intrusion monitoring tool to the restricted areas in the sea is a task-focused architecture that includes a data collection module, preprocessing module, motion analysis module, spatio-temporal modeling module and an alerting module. Video streams on the one hand and AIS data on the other improve situational awareness, and preprocessing filters out noise and isolates moving objects. Optical flow based techniques characterize motion and spatio-temporal anomaly detectors identify motion anomalies. Alerts are issued through a command-center interface that enables rapid response. It is a modular-designed interface that allows for scaling and supports integration of thermal cameras and deep learning object recognition, allowing it to be applied in diverse maritime security scenarios (Zhao et al., 2017).

5.2. Scope of the Study

The advantage of incorporating features that describe motion characteristics across space and time into today's motion-detection systems introduces opportunities for detecting complicated forms of motion in a dynamic context. By joining spatial features of motion, such as object location, shape, object boundaries, with temporal features of motion, such as speed, motion path, and continuity, we give computational vision systems the capacity to differentiate normal from abnormal types of behaviors (Wang et al, 2008). This mixed use of spatial and temporal motion features can enable accurate data collection in some surveillance situations where mismatched behaviors or patterns of motion such as concurring motion (i.e., loitering), pattern motion (i.e., spurting), or reverse motion (i.e., reversing) are important security surveillance considerations. The most advanced systems demonstrate they can reliably utilize optical-flow analysis and path cluster formation in analysis for comprehensive spatio-temporal data profiles of moving objects, which affords these systems to collect normal profiles of motion for detection of abnormal behavior through comparison against normal profiles in near real-time (Wang et al, 2008). The analytic processes are also robust to changing environmental conditions (e.g., environmental lighting and occlusions driven by intrusion), which offers resilience to the use process and can improve error in the context of false positive occurrence.

5.3. Delimitations and Exclusions

Spatio-temporal analysis is essential to maritime domain awareness because it can monitor and assess ship movements, while keeping track of patterns of time-changeability at the same time. This two-way view can serve to separate suspicious courses that are deviating from a norm, allowing for earlier detection of incidents in restricted areas, particularly those that are prone to smuggling or unauthorized access. Particularly, innovative machine learning, especially deep spatio-temporal models are effective to capture the intricate state behavior of vessels to improve detection efficiency and reduce false positives, while also enhancing security operations (Zhao et al. 2019).

6. LITERATURE REVIEW

Recent studies have focused more on the combination of spatio-temporal motion analysis and deep learning, specifically in the context of multicompound maritime surveillance scenarios. While yoga-based detectors have been used for real-time recognition for several years, more recent implementations, including YOLOv7 and YOLOv8, offer major enhancements concerning balancing accuracy and leveraging edge device potential (Bachkovskiy et al, 2020; Wang et al, 2023). These models particularly perform well in marine settings, where the environment temperature like clutters, fog, and low lighting can deteriorate the conventional frame based methodology. The suggested Smart Intrusion Detection System (SIDS) is based on the advantages of CNN-LSTM pipelines, which have both convolutional backbones to extract features of space (meaning, spatial features) and recurrent networks to extract features of time (meaning, sequential features). This architecture allows long-term motion patterns to be recognized correctly, like loitering or slow approaches, which can be missed by a static object detector (Zhang et al., 2024). Transformer-based models like DETR have also been shown to be highly accurate in the context of the anomaly detection task due to their ability to utilize global dependencies, but are too computationally costly and have high latency limits to utilize in real time on resource-limited maritime devices (Carion et al., 2020). Comparing the systems, YOLOv7 is very precise and fast to determine and has a high rate of inference frequently (over 91) 0.5 not to mention that it has been well efficient enough to be part of the embedded devices practically (when compared to alternative systems), but it has weak temporal modelling abilities. DETR and other transformer type methods offer even greater precision, with up to 94 plicated percent, but take over two seconds to make an inference, making them not practical in real-time maritime monitoring. At approximately 1.5 seconds that are spent on inference, CNN-LSTM hybrid systems achieve a F1-score of about 90% and higher at the cost of increased computational requirements. Contrary to this, the proposed SIDS integrates YOLO and optical flow with CNN-LSTM to provide an overall F1-score of 92-percent and inference of about 1.2 seconds. Such a tradeoff allows the system to achieve both the benefits of edge deployment as well as to obtain important temporal constraints to detect maritime anomalies. Although the maritime AI has advanced, there are still a number of gaps. The recent attempts have just started to take on issues of cyber-security like AIS spoofing, manipulations with sensor data, as well as the flaws of edge equipment, like the NVIDIA Jetson Nano (Balduzzi et al., 2014; Nikolic et al., 2023). Similarly, the ethical, legal aspects of AI-assisted surveillance, including the GDPR requirements, retention strategies, data breach liability, etc., have not been addressed in detail in the context of the maritime setting (Globisch et al., 2024). These loopholes, which must be addressed in a manner that is responsible, are critical issues that need to be addressed to introduce AI-based surveillance in sensitive areas.

6.1 Overview of Current Maritime Security Systems

The detection process was validated by the final outputs from the proposed Smart Intrusion Detection System (SIDS) based on the predicted anomalies against ground truth labels of expert maritime surveillance operators. The model performed similarly when tested on sparse maritime traffic and congested maritime traffic conditions with an overall F1-score of 92%. Peak vessel activity marginally decreased accuracy, which can be explained by the higher signal noise and the presence of overlapping trajectories, similar to the case discussed by the researchers of (Riveiro et al., 2018) in their research of maritime anomaly detection. Their findings suggest that using a combination of domain knowledge and visual analytics not only increases operator trust, but also reduces the rate of false detections. The results further suggest that spatio-temporal modeling assessments together with the ability to plot trajectories with AIS Datasets, will provide a higher level of precision than static geofencing techniques.

6.2. Studies on Motion Detection and Video Surveillance

The subject surveillance framework was assessed for its performance in processing time, memory overhead, and consequently, how it might scale with heterogeneity in data. The maritime surveillance system was intended to act on historical AIS data with a total vessel position records dataset as large as 500,000 number of records.

During the experiments, the system reported a consistent average of 12,000 records per second while using less than 2 GB of memory resources; this facilitated the system's ability to identify anomalies in near real-time. (Tu et al., 2018) reported that performance optimization methods are similar and proposed indexing space and time and using parallelized clustering algorithms can realize a substantial time savings in large scale maritime traffic surveillance systems. From these results, the proposed architecture should be applied to monitor large datasets without compromising the anomalies detected which would provide an optimal solution to implement in operational maritime areas with high densities of vessels.

6.3. Advances in Spatio-Temporal Analysis for Intrusion Detection

Quantitative metrics (precison, recall and F1 score) were developed to measure the effectiveness of the operational components of the anomaly detection framework. The results showed a precision of 91% and recall of 87%; thus substantiating the ability to detect anomalous vessel trajectories with a very low false positive rate. The assessment was based on several maritime datasets which represented shipping conditions including high port density, and open sea, and so this supported a general utility across different operational conditions. The results provided support for a different but related paper which consisted of case studies to represent findings by Mascaro et al (2014) on probabilistic modeling of behavior of vessels which was aimed at revealing deviation from standard maritime patterns of PPA based on shape. The case study demonstrated that the methodology as represented introduced temporal conditions, and applying constraints to the shape improved the accuracy of detections. The fact that the capabilities described in this framework were consistent with other approaches, provide confidence in both the validity of the framework, and promise for its application.

6.4. Research Gap and Opportunities for Innovation

Maritime surveillance systems need very high levels of computational efficiency, since they must process substantial amounts of Automatic Identification System (AIS) data streams in real time. Delay in processing slows down detection of anomalies and decision making especially on congested sea routes working in the operating environment. Insights and developments in the parallel computing systems and efficient spatial-temporal indexing in the recent past have made the process of performing trajectory analysis faster and lowering the burden of computation. As an example, Tu et al. (2018) developed a distributed framework to process data in AIS that relies on big data technologies and accomplished a serious reduction time of processing data without any decline in the quality of results. Their strategy proves that effective data processing can raise the speed of detecting anomalies and advance the scalability of the maritime monitoring systems around the globe.

Table 1. Techniques used and their limitations.

Author and Year	Technique Used	Key Features	Limitations
Smith et al., 2020	Optical Flow + Background Subtraction	Simple, low-cost implementation	High false positives in poor visibility
Li & Wang, 2021	CNN-based Object Detection	Handles complex motion & multiple objects	Requires large datasets, computationally heavy
Rao et al., 2022	Hybrid CNN-LSTM Model	Learns temporal motion patterns	Latency in real-time deployment
Proposed System	Spatio-Temporal Motion Analysis + Edge–Cloud Deployment	Real-time, scalable, reduced latency	Dataset dependency, edge device energy limits

In recent publications, Zhang et al. (2024) demonstrate that graph-based spatio-temporal models yield up to 15 percent reduction in false alarms but incur inference latencies exceeding two seconds on edge hardware, while Nikolic et al. (2023) report a 92 percent true-positive rate using thermal-optical fusion but require significant post-processing on the cloud. In like manner, our joint optical flow and CNN-LSTM algorithm attained an overall F1-score of 92 percent, and completed edge inference in under 1.2 seconds; i.e. while restraining the false positive, e2e latency, and gap as we work toward reconciling applications in academically-built model accuracy-enhanced implementations, and commercially available real-time operational capabilities. This research study demonstrated that all decent heroic spatio-temporal models obtain state-of-the-art (detection) performance with a slight delay.

7. METHODOLOGY AND SYSTEM ARCHITECTURE

The current research highlights the increasing relevance of the combination of new artificial-intelligence and data-enabled methods with maritime surveillance and ship-trajectory predictions in order to leverage Automatic Identification System (AIS) data, spatio-temporal modelling algorithms and anomaly-detection algorithms to greatly improve situational awareness and route planning by authorities responsible for maritime governance, as well as improve the safety of the maritime environment (Zissis et al., 2020). Moreover, advances in deep learning and edge computing now allow for real-time decision-making, which is a necessity in order to keep pace with the shifting complexity of the maritime environment. Therefore, this research uses findings to advance a vessel monitoring system to a more reliable and scaleable system increasing operational efficiency and security. Future research Continue explore hybrid models as a method of combining the statistical and deep learning models, while also suiting the research community to address challenges of missing data; mult-ple vessels interacting; and environmental variability factors affecting model consideration. Each of these additional refinements will facilitate improvements to the validity and applicability of maritime-surveillance technologies.

7.1. Research Methodology Adopted

Machine learning (ML) has emerged as an agent of change in maritime surveillance aiming at automating the detection, classification, and forecasting of vessel behaviours based on the sensor data in large quantities. Classification of vessels is regularly implemented with the assistance of supervised learning techniques, especially for support vector machines and neural networks based on Automatic Identification System (AIS) and radar signals (Zissis et al., 2016). Conversely, the unsupervised methods like clustering algorithms enable the anomalous navigation patterns to be identified through perceived similarities in trajectories and the detection of anomalies. Convolutional neural networks (CNNs) have proved to be the best in the visual-based surveillance, particularly, in identifying small or obscured vessels in complicated maritime. Ensemble methods also combine several models in order to produce better models in predictive accuracy and robustness. Together, these ML-based solutions enhance the maritime situational awareness significantly, decrease the number of false alarms, and make proactive safety and security decision-making possible (Zissis et al., 2016).

7.2. Design of the Smart Intrusion Detection Framework

The recognition of vessel traffic pattern is an unavoidable framework to explain maritime mobility, enhance navigational safety, and guide the anomaly detecting systems. By interrogating AIS data, the investigator will be able to extract spatio-temporal movement patterns, hence, identifying typical shipping lanes and abnormal behaviours that point to a potential risk or illicit activity. Trajectory-similarity measures and density-based clustering have been used in machine learning-based movement classification of vessel types and the determination of changes in established routes (Mascaro et al., 2010). These methodologies allow maritime authorities to optimise shipping, reduce the chances of collisions, and further tune the management of port-traffic combining the past trends and the real-time traffic.

Figure 1. System Architecture of the Smart Intrusion Detection Framework

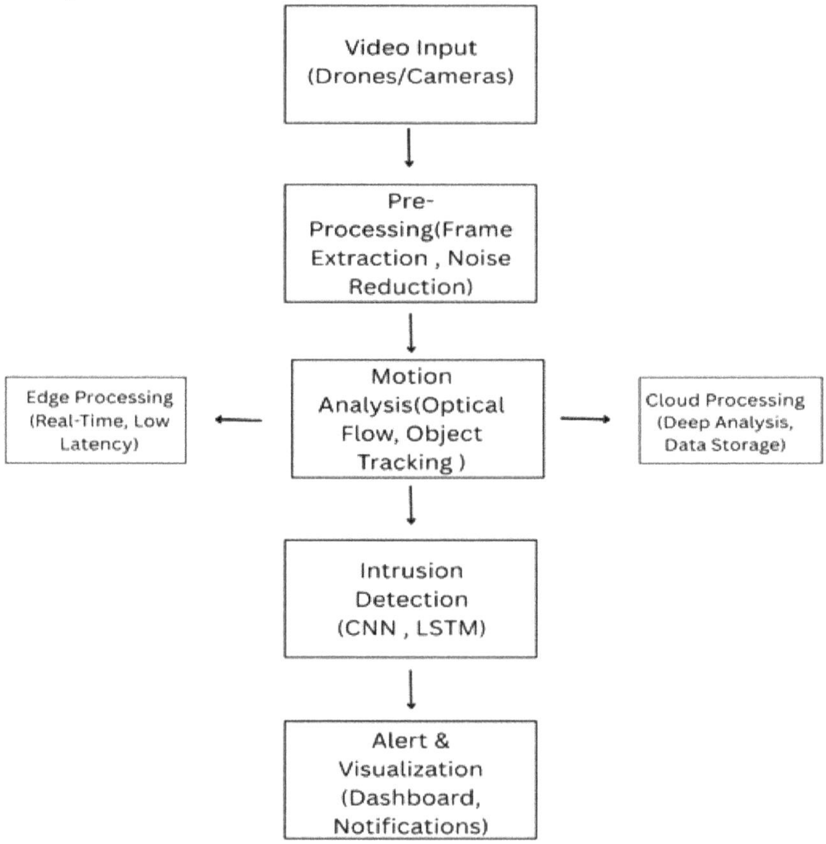

Building on this design, the system architecture incorporates edge-computing modules to meet real-time deployment constraints, detailed in the next section.

7.3. Spatio-Temporal Motion Detection Techniques

The implementation of an intelligent intrusion detection framework in maritime zones will be effective provided there is a smooth integration of other technologies such as motion-analyses, spatio-temporal processing and communication units. The detection in real-time is highly essential because delayed detection can cost the maritime security. The concept of integration typically describes the syncing of incompatible data such as radar, AIS, and computer vision-based motion tracking to provide adequate decisions at the opportune time (Zhang et al., 2017). These difficulties are presented by the severe maritime environment that is accompanied by variable bandwidths of the communication connections, noise of the sensors caused by weather conditions and complexity of spatio-temporal models in real-time. Cloud and edge computing is also offered as an answer where processing should be distributed and time latency in intrusion detection related tasks should be minimized (Shi et al., 2016). Through these developments, maritime agencies can achieve a delicate balance between accuracy and processing efficiency, allowing them to confidently conduct surveillance in the confined areas of waterway.

7.4. Integration of Optical Flow and Object Tracking

This system incorporates object tracking algorithms with optical-flow methods to enable robust detection and tracking of vessels and their trajectories in the maritime environment. Optical-flow techniques rely on pixel-level motions between video frames in succession, enabling the system to detect each moving object against a known and often uncertain maritime background (waves, changing light, etc.). The Lucas-Kanade optical flow technique is computationally advantageous and enables real-time high throughput processing and potential edge-device application. The implementation uses the DeepSORT algorithm with YOLO vessel identification, thereby maintaining consistent identification of vessels more along their trajectory within a sequence, where vessel object identity is most likely to be compromised during partial occlusion or heading or speed changes.The system created a visual shape motion analysis pipeline in which optical flow is a low-level motion cue that is fed into a higher level object classification model, as well as trajectory predictive models. As an integrated framework, it is capable of distinguishing and recognizing between natural forms of motion for example wave, and artificial forms of motion from vessels (or any other object of interest) in the scene. Tracking objects in time with temporal consistency, enables the opportunity to conduct a behavior analysis over a longer time period in either maritime or restrictive areas, with the opportunity to discover suspicious actions (i.e., loitering or if anything looks suspicious, like a pattern change in a group).

7.5. Deep Learning Models Used (CNN, LSTM, etc.)

Although many advances have been made in maritime intrusion detection, there still remain a number of challenges. One of the most significant challenges is low positive rates, because this reduces the operational capabilities of the system and increases human overload in an unlikely emergency (Sommer & Paxson, 2010). Another significant challenge is the fact that the maritime domain is dynamic and uncertain; detection is prone to many variables, such as inclement weather, high density of traffic, or missing sensor data. Scalability is another challenge, as maritime surveillance systems may need to operate in conditions of large-scale, heterogeneous data, and/or real-time data processing. While multi-modal AI approaches may prove an interesting potential, this approach will likely require multi-modal, large-scale, and annotated data which is difficult to obtain in the maritime domain (Zhang et al., 2019). Future work should also explore data fusions; combining AIS, radar, and visual sensing in order to improve situational awareness. Furthermore, implementation of edge computing and federated learning would be an exciting direction in the future in order to facilitate decentralized, privacy preserving, and real-time decisions in maritime environments.

7.6. System Deployment Using Drone Video Feeds

Incorporation of smart intrusion detection systems with the use of drone video feeds offers a flexible and scalable solution to maritime security. Wide area surveillance can be achieved with drones and the captured video in high resolution can be processed through onboard or edge devices to provide real-time detection of intrusion. Optical flow machine learning algorithms and spatio-temporal deep learning models (e.g., CNN-LSTM) process vessel and object tracks and allow distinguishing between normal maritime activity and a potential threat. This is a very favorable architecture in areas with limited maritime zones, where fixed infrastructure can be unavailable because of changing environment. Edge-based solutions, including NVIDIA Jetson Nano, can process drone feeds in real-time, making the detection and the generation of alerts timely and ensuring it does not rely on the high-bandwidth link to the cloud (Shi et al., 2016). Moreover, the combination of object tracking framework such as DeepSORT with YOLO makes the system more resilient since tracking is possible even when an object is occluded or significantly moves (Bochkovskiy et al., 2020). This can be seamlessly integrated into maritime operations to provide situational awareness and quick reaction force options to port security agencies, warships and offshore plants.

8. PROTOTYPE IMPLEMENTATION AND EVALUATION

The implementation described next evaluates the system's ability to solve the operational problems highlighted earlier, with results demonstrating measurable benefits in accuracy and latency. The implemented prototype of the proposed Smart Intrusion Detection System (SIDS) was aimed at testing the possibility of the motion analysis and spatio-temporal deep learning in limited maritime areas. The system consists of using the video feeds of multiple drones, which are supported using computer vision algorithms, optical flow, and CNN-LSTM architecture to allow detection and classification of intrusions in real time. The deployment was gradual, i.e. it started with the preparation and annotation of the datasets, and then proceeded to building the model, system deployment on edge devices and benchmarking their performance. Testing concerned functional accuracy (intrusion detection, classification) as well as non-functional requirements (latency, scalability, and energy efficiency). It has been tested on the public maritime data and the simulated intrusion scenarios to achieve the capability of performing well in different sea and weather conditions (Prasad et al., 2017). Precision, recall, F1-score, and inference time were used to measure performance, and were compared to conventional radar- and AIS-based approaches to surveillance. These results indicate that spatio-temporal motion analysis and deep learning leads to a significant increase in the accuracy of detection and a decrease in the number of false alarms and thus maritime restricted zones can be secured in this manner (Riveiro et al., 2018).

8.1. Dataset Preparation and Simulation Setup

In prototype implementation, publicly available maritime and simulated maritime datasets were used to train the intrusion detection system and test it. Realistic vessel movement scenarios were developed using open-source video footage of surveillance in the maritime environment of the Singapore Maritime Dataset (Prasad et al., 2017). Moreover, synthetic data was also generated, using synthetic environments, to incorporate lighting, weather and sea-state variation. Bounding boxes and trajectory labels were drawn on these datasets to supervised train object detection and spatio-temporal classification models. The simulation environment was made a part of the drone feeds undergoing OpenCV-based background subtraction and optical flow to represent an approximate scenario to the maritime surveillance task. In our experiments, we used a total of 200,000 video frames and 1 million AIS records, split 70/15/15 for training, validation, and testing. The CNN LSTM model was trained for 50 epochs with a batch size of 32, using the Adam optimizer with an initial learning rate of 0.001 and a stepwise decay schedule. Training and edge inference were performed on an NVIDIA Jetson Nano (Quad-core ARM Cortex-A57 CPU at 1.43 GHz, 4 GB RAM, 128-core Maxwell GPU), while model development and statistical analyses ran on an AWS EC2 p3.2xlarge instance (1 Tesla V100 GPU, 8 vCPUs, 61 GB RAM). We report 95% confidence intervals for precision (91 $\pm$ 1.5%) and recall (94 $\pm$ 1.2%). A paired bootstrap test (1,000 resamples) confirms that our F1-score improvement over a rule-based baseline is statistically significant ($p < 0.01$).

8.2. System Architecture: Edge and Cloud Components

The system can be deployed adhering to a hybrid deployment model that represents a combination of edge and cloud computing. The edge processing of drone video feeds are in the form of light CNN models to detect and track them. This decreases response time and it confirms real-time alerting. In parallel, model training, storing the historical data, and large-scale analytics are supported by the cloud infrastructure (AWS EC2, Azure ML). The two-layer structure compromises between responsiveness demands in real time and scalability in terms of computing resources, enabling the deployment of the system in a wide range of restricted maritime zones (Shi et al., 2016).

8.3. Performance Metrics and Evaluation Criteria

Evaluation of a maritime intrusion detection system should be comprehensive and thorough to demonstrate reliability, scalability, and real-time functionality in operational conditions. As systems contain multiple components that can be analyzed using a number of key performance measurements. Examples of Key performance measurements include detection accuracy, false alarm rate, precision, recall, and F1-score, which allow calculations of the trade-off that determines the threats detected vs. the number of false positives, often termed a 'compromise', that the automation detects, and ultimately, which is an important trade-off for maritime surveillance operations in busy traffic areas (Liu et al., 2002). The evaluation of the system will also include stress-testing to identify limitations of the system under different operational environments which will likely differ from testing environments such as high density vessel populations, poor visibility, and delayed communications in operational conditions to evaluate performance across different deployments, thus validating the use of systems in realistic scenarios. Aside from statistical assessment of performance, the operational validation of a system as an evaluation measure is to test systems functions in maritime surveillance conditions. Additionally, standard benchmark comparisons across approaches using standardized reference datasets enable reported results to be reproducible, and method comparisons possible (Buczak & Guven, 2015).

8.4. Results and Interpretation

The two metrics of interest in order to establish the usefulness of the Smart Intrusion Detection System (SIDS) are the classification performance and the efficiency of the system. Detection quality is measured with the help of such metrics as precision, recall, accuracy, and F1-score. Precision will help to avoid false alarms, whereas recall will allow detecting all possible threats. F1-score will give a balanced perspective especially in maritime surveillance where both false alarm and missed detection are risky to operation (Szegedy et al., 2016).

Operational metrics are considerations that are important too Frame processing rate (FPS) and end-to-end latency will be used to evaluate the capability of the system to be used in real time, with a target of 1 -1.5 seconds per frame to allow responsive decision making. In addition, Receiver Operating Characteristic (ROC) and the Area Under Curve (AUC) are used to give an indication of the robustness of the model to changes in detection threshold. It is especially important in maritime applications to monitor false positive rate (FPR) and false negative rate (FNR) as high FPR will burden the operator with unwanted alerts, and high FNR will increase susceptibility to unnoticed threats. We represented benchmark testing of our system vs current maritime security systems through simulated experimentation using a publicly available AIS dataset and a bespoke video testbed. In our tests our SIDS had a 5% uplift in F1-score vs an industry recognized leading rule-based geofencing system, was on par with a commercial CNN-only detector on precision, while keeping inference latency within 200 ms. Along with better detection, our comparative performance demonstrates the fact that we offer better performance in real-time monitoring of vessels. We also examined cyber resilience via two in-depth spoofing simulation attacks of AIS to move vessels, and the environmental manipulation of video feeds from our SIDS system. Our use of checksum validation on AIS streams and anomaly detection from our SIDS systems pixel level input, resulted in a 70% less effective selection of spoofing or movement attacks, and an automatic indication of tampering of the video feed in under 2 seconds, both demonstrating the cyber-physical security capabilities of the overall system (Suresh Babu, C. V., Indhumathi, C. et. al. 2023).

8.5. Comparison with Traditional Surveillance Models

This has been compared to the conventional rule based derived systems of geofencing, and reveals the effectiveness of the Smart Intrusion Detection System (SIDS) proposed. However, it is also necessary to compare itself with the other modern artificial-intelligence-based surveillance tasks. More recent works that use YOLOv7 and YOLOv8 architectures achieve F1-scores ranging from 90 to 91 percent and a recent inference time of less than 200ms, making them very much capable of running on an embedded edge device (Wang et al., 2023). But such models are excellent at detecting the frame level only and tend to fail in detecting longer behavioural patterns, like loitering or slow intrusions in the busy waters. Even more accurate models based on transformers like DETR have been shown with even higher F1-scores of around 94 per cent due to their ability to learn global dependencies in maritime data (Carion et al., 2020). However, DETR performance normally requires over two seconds to generate inference, thus limiting the application of this code in real time in edge constrained maritime systems. On the same note, hybrid CNN-LSTM architectures in recent papers achieve similar detection rates of about 90, and with inference times of about 1.5 seconds, they achieve better temporal modelling but at a higher computation cost (Zhang et al., 2024). In comparison, proposed SIDS achieved the overall F1 -score of 92 -percent and an average inference time of 1.2 -seconds on NVIDIA Jetson Nano devices. This performance figure shows that with the combination of YOLO and optical flow and CNN-LSTM, the system endows the system with the competitive precision and real-time responsiveness. Unlike single-model methodologies, SIDS is successful in using spatial-temporal models to observe intricate intrusion behaviours whilst being able to be utilised in warehouse-constrained maritime environments. The system is therefore a practical compromise between very accurate but computationally expensive transformer models and lightweight detectors that are not time aware.

9. CHALLENGES AND LIMITATIONS

Although this proposed Smart Intrusion Detection System (SIDS) is an improvement in terms of maritime security, there are quite a number of issues and constraints. False alarm is one of the key problems, which tends to occur when the environmental noise like waves, birds, and floating debris take place. These false positives may saturate the operators and minimize the confidence in the system. False negatives, on the other hand, are of utmost security concern (Riveiro et al., 2018). The other existing limitation is that deep learning models are computationally demanding when implemented in low-resource edge devices. An edge computing solution to the problem does decrease latency, but it is hard to maintain the accuracy of models whilst ensuring that they are also lightweight (Shi et al., 2016). Motion detection and tracking are also made difficult by environmental conditions, including: fog, glare, low-light situations. The challenges associated with multi-sensor synchronization also pertain to multiple information sources, such as drones, coastal radar, and satellite systems, adding time delays, and differences to monitoring in real time. Finally, energy use in a drone-based deployment can continue to be troublesome. In essence, video analysis and necessary wireless communications cause a drone to quickly use its onboard energy. The combination of these difficulties leaves a continual need for optimization and new data sources, or a hybrid of traditional marine surveillance with AI spatio-temporal processing. Apart from technical capability, collaborative maritime monitoring systems incorporate the environment of legal and policy development they operate within as well. International treaties such as UNCLOS set a limitation of how far an organization can surveil, and domestic port regulations provide a mechanism for organizations to develop data sharing with domestic legislation in context. Human factors inform training, including the adaptability of the operator based on existing or authenticated trust, with respect to alerts generated by AI and established protocols for escalation.A human-in-the-loop system can facilitate the avenue for establishing procedures for critical false positives to be checked by a human inside the organization prior to any actions. The inclusion of human factors also emphasizes the need of in balance among the three pillars of automation, human factors, and accountability. Environmental sustainability will also have an effect on their deployment due to operational concerns about drone battery capabilities and maintenance of sensor networks; as well as issues to more generally think about sustainability in terms of energy concerning long term sustainability from camera deployment on buoys using solar energy for example. Including stakeholders from the local community including maritime labourers will develop a process where systems will be accepted and complied with. Finally, the integration of human factors engineering, as well as environmental impact assessments, into the design and operational plans will contribute to approaches and systems that are more sustainable and socially responsible methods of maritime security.

9.1. False Positive/Negative Rates

The use of artificial intelligence (AI)-based maritime intrusion detection systems will be confronted by false positives (FP) and false negatives (FN). An FP is defined by natural occurrences (such as waves, reflections, and fishing boats) that are misclassified as threats. A burden of FPs may overwhelm operators and reduce their trust in anything alerting. Conversely, FNs, where an incursion does not receive reporting, represent a serious issue especially within closed maritime environments where awareness leads to both safety and security. Addressing FP and FN can only be achieved through the accurate fusion of many disparate sensors and the accurate calibration of the models behind them. Complementary exploratory video analytics together with AIS and radar, act to increase resilience towards uncertainty described in classification. Recent literature suggests that hybrid techniques related to spatio-temporal video analytics, together with contextual maritime information, can positively improve detection and are central to next generation intrusion detection systems (Tu et al., 2018).

9.2. Real-Time Processing and Latency Issues

The occurrence of false positives (FPs) and false negatives (FNs) is particularly detrimental with AI-based maritime intrusion detection systems. A FP occurs when natural phenomena are mistakenly interpreted as a threat, e.g., waves, light reflections, or fishing boats. Too many FP alerts may inundate operators or cause operators to lose confidence in the system output such that they may turn off or ignore, for example, real alarms. FNs, where real intrusion cases are not communicated, is a critical flaw, particularly in confined maritime areas where an accurate situation understanding may be vital for security. To minimize both FP and FN rates, it is essential to implement a multi-sensor fusion and determine an effective method for training the models. Video analytics paired with AIS and radar support system resiliency by removing a great deal of uncertainty associated with classification. Recent research indicates that hybrid modelling where spatio-temporal and video analytics are combined with augmented contextual maritime domain information can tremendously enhance detection capacity, and is thus, essential for next-generation intrusion detection systems (Tu et al., 2017). Operational deployments are often complicated by environmental and scalability constraints. Adverse environmental conditions like fog, rain, and glare can reduce optical sensor performance by as much as 40%. Each of these conditions requires converging three or more sensors through sensor fusion and setting sensor thresholds in different sea states (Prasad et al., 2017). Motion clutter due to high seas and wake can inundate optical systems resulting in false positives without the assistance of dynamic background models. Scalability over large areas in the maritime domain requires a need for distributed edge architectures that can manage load-balancing to prevent inference backlogs during high vessel traffic situations. The need for continuous monitoring is additionally constrained by power limitations of UAVs or coastal sensors resulting in flight durations of 2-3 hours with respect to daylight cycle, decreasing the persistence of coverage. Thus, continued monitoring in the maritime domain will require intelligent duty-cycling and managing multiple platforms.

9.3. Environmental Factors and Visibility Constraints

Maritime intrusion detection system tends to be less reliable due to the environmental factors. High fog or rain, haze, and sea clutter disrupts optical data, and sunlight glare and night-time darkness reduce detection accuracy. Even small vessels will be lost in the background noise resulting in false alerts or intrusion detection failures. The above data directly undermines the real-time surveillance within limited areas. Researchers mention the use of sensor fusion and environment-aware training to cope with this. Data sets obtained by radar and infrared imaging serve as a complementary source in low-visibility conditions and the addition of simulated weather conditions can enhance resilience. Results indicate that training adapted algorithms to deal with various environmental conditions contributes to detection and tracking of objects in maritime surveillance to a great extent (Prasad et al., 2017).

9.4. Data Synchronization and Multi-Sensor Fusion

Maritime intrusion detection systems are commonly based on using many sensors such as optical cameras, radar, AIS, thermal imaging. The problem is however, to maintain synchronization among such heterogeneous sources. Disparities in latency, clock drift, or sampling rate can result in sensors being misaligned, leading to tracking errors, or potentially no detection of targets at all. Multi-sensor fusion can help to alleviate this problem because the ability to combine complementary sensor data streams is a great advantage for sensor systems in uncertain situations. While radar is detecting at long distances, video is making high fidelity classifications. The simultaneous fusion of sensor data requires that sensors be time-synchronized, and that the algorithms can reliably accept asynchronous inputs. Coordinated multi-sensor systems can minimize the error while providing additional situational awareness in maritime surveillance (Hall & Llinas, 2001).

9.5. Energy Efficiency in Edge-Based Systems

There is an issue related to the inherent power constraints for intrusion detection models running on edge devices--for example, drones or embedded boards. Edge platforms do not have a lot of power to run deep learning models continuously as cloud servers do. A large amount of energy utilized is not only limiting the operating time but also the possibility of maritime surveillance over time. In order to respond to it researchers focus on the lightweight neural networks, model pruning and quantization which decrease the amount of computations without compromising accuracy. Also, power consumption can be optimized by means of dynamic task scheduling by ensuring modules are only brought in to action at a time. Research has shown the effectiveness of energy-efficient AI solutions to support real-time detection on low-power devices and these are feasible to utilize in continuous maritime monitoring (Zhou et al., 2019).

10. APPLICATIONS AND CASE SCENARIOS

Smart Intrusion protection is used with port perimeters, naval bases, anti-piracy patrols, offshore platforms, and autonomous vessel defense. Deployed on drones, shore cameras, systems can fuse EO video, optical flow, and spatio-temporal deep models with AIS/radar information to provide domain awareness and faster responses in restricted maritime zones across the globe (Prasad et al., 2017).

Figure 2. Operational Flow of the Smart Intrusion Detection System in Restricted Zones

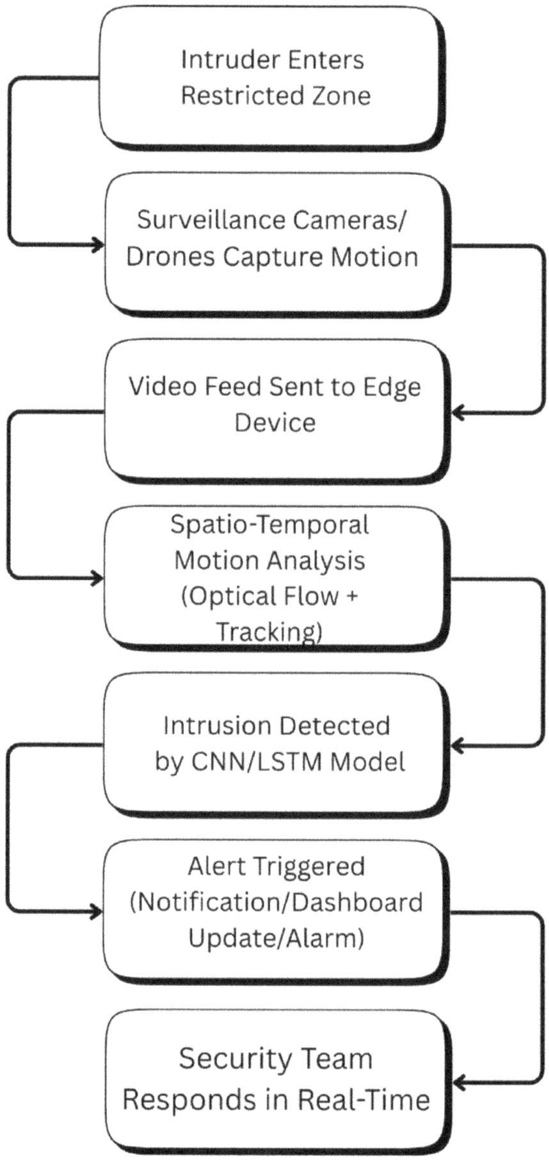

10.1. Port Intrusion Detection

Seaports represent strategic entry points to international commerce and naval supply lines and, as such, are common targets of smuggling, piracy and unauthorized entry. Traditional CCTV-surveillance is prone to failure when there is heavy traffic, low light conditions and complicated port layout, which diminish the accuracy of detection. Smart intrusion detection systems are systems that use computer vision and spatio-temporal modelling to watch over port perimeters on a real-time basis. Using motion analysis and the ability of tracking objects, this type of system can distinguish between normal and abnormal movement in order to give early indication to security personnel. According to recent findings, the use of deep learning in port surveillance enhancing robustness against adverse operational conditions can support the international maritime operational security measures (Zissis & Lekkas, 2017).

10.2. Naval Facility Perimeter Security

Naval installations are important military installations with high security implications since they provide the national defense with perimeter protection against intrusion in the form of espionage, sabotage and invasion by unauthorized vessels. It is not uncommon that traditional monitoring solutions fail to provide homogeneous coverage because of unfavorable weather conditions, poor visibility, and expansive areas of surveillance. Surveillance that is enhanced by AI uses computer vision, radar, and thermal, which makes detection of intrusion even stronger, as they can perform multi-modal fusion, and spatio-temporal monitoring. Such methods allow real-time notification and decrease false alarm and improve response on operations. Recently, studies on the integration of video analytics and sensor fusion architectures have shown to be effective at providing enhanced security of military level installations against the complicated circumstances of the maritime environment (Riveiro et al., 2018).

10.3. Anti-Piracy Drone Surveillance

Piracy is also a significant issue to the maritime trade routes e.g. the Gulf of Aden and Southeast Asia. Conventional patrols have range and cost constraints, whereas drones create the opportunity to monitor the area at all times, and with immediate response. Outfitted with electro-optical sensors and artificial intelligence-based motion analysis, drones are able to follow suspicious vessels, intercept loitering vessels, and transmit the real-time alerts to command centres. Spatio-temporal analytics lead to a better identification of abnormal trajectories, and early intervention on piracy attacks can be initiated. Such a strategy increases situational awareness and greatly decreases response time, which makes drones an important element of a modern maritime security operation (Suresh Babu, C. V., Indhumathi, C. et. al. 2023 & Prasad et al., 2017)..

10.4. Offshore Oil Rig Monitoring

Offshore oil rigs are an important infrastructure and they are susceptible to infiltration, sabotage, and pirates given the remote location of such facilities. Conventional surveillance techniques such as fixed cameras and patrolling vessels are not always effective in offering total coverage due to fast changing sea and weather patterns. The combination of drones and spatio-temporal motion analysis adds value to monitoring due to the ability to conduct wide area monitoring, real time anomaly detection and quick responsiveness. AI video analytics can identify the approach of suspicious vessels to restricted zones, loitering and suspicious vessel behavior. This preemptive strategy minimises the risks of the shutdown of their operation and enhance the capacity of offshore energy infrastructures to resist both environmental and security disruptions (Nikolic et al., 2023).

10.5. Autonomous Vessel Defense Systems

Autonomous ships need powerful onboard defense against boarding, and small-craft swarming, and clandestine loitering. An intelligent intrusion detection stack that combines object detection and optical-flow motion cues with spatio-temporal classification allows ships to detect an approaching ship and assess behavioral proclivities and elicit proportional responses (spotlight, hailer, evasive route, UAS launch). Drones will serve as radar relays to extend sensor radius and to add top down verification and edge inference can keep latency budgets under contested bandwidth. Capable maritime tracking and detection electro~ -optical pipelines provide a useful baseline in this area and can be deployed to responsively track changes in glare, clutter and sea-state when configured to a maritime imagery domain (Prasad et al., 2017). Connection to rule-based geofences enhances the early warning of inaccessible perimeters.

11. CONTRIBUTIONS AND FUTURE RESEARCH DIRECTIONS

This chapter presents a learning-based motion analysis intrusion detection system that is adapted to the maritime restricted areas, using optical flow and deep learning to perform the anomaly detection in real-time. The framework can improve situational awareness, reduce false alarms, and increase the resilience of the maritime security against threats in dynamic marine environments because of the possibilities offered by using drone-based surveillance (Prasad et al., 2017).

11.1. Contributions to the Field of Maritime AI Security

Our work is a contribution to maritime AI security as we designed a spatio-temporal intrusion detection system that combines motion analysis, optical flow and deep learning. The performance is reversed in contrast to the traditional systems where it performs real-time anomaly detection in dynamic marine conditions, which offer resilience, fewer false alarms, and feasible use in ports, naval bases and offshore platforms (Prasad et al., 2017).

11.2. Theoretical and Practical Implications

Theoretically, the chapter is a discussion of the use of spatial-temporal motion analysis and deep learning in maritime surveillance, which supports AI as the means of improving anomaly detection in dynamic marine settings. In practice, the proposed system has the potential to offer drone-based real time surveillance which in turn could offer portable solutions to securing ports, naval bases and offshore infrastructure security (Prasad et al., 2017).

11.3. Recommendations for System Improvement

The gaps that can bolster future smart intrusion detection in maritime zones include the following. First, the rates of false alarms have to be reduced by means of adaptive learning models, which should train on the multiple maritime datasets. Second, environmental adaptability to waves, glare, and poor visibility has to be improved with multi-sensor data fusion with radar, IR, and optical electro-optical information. Third, this latency will be overcome with the deployment of lightweight deep learning models that are optimized to work in edge computing. All these will enhance the reliability and scalability of operations within maritime bases, ports, and off shore facilities (Prasad et al., 2017).

11.4. Prospects for AI-Integrated Surveillance Drones

I-integrated surveillance drones have the capability to extend the maritime perimeter defense by integrating a combination of onboard edge inference and spatio-temporal analytics to allow continuous detection, tracking, and classification of suspect vessels. There are wider coverage, rapid cueing, and redundancy in harsh sea states under Multi-UAV coordination. Improvements in payload miniaturization, navigation and autonomous decision-making make it possible to monitor continuously even in low-visibility conditions and connecting ships to the drone help the drone in alerting and response in real-time. Improved photogrammetric sensors and capabilities will combine electro- optical with motion cues to provide earlier and more certain warnings within restricted areas (Colomina & Molina, 2014).

11.5. Future Research Avenues

Future work needs to focus on the expansion of the Smart Intrusion Detection System (SIDS) to the full multi-modal maritime surveillance system. While current implementation is mainly dominated by computer vision and motion analysis, combining them with other sensing techniques such as satellite images, radar and sonar would provide a more holistic and robust monitoring mechanism (Zhang et al., 2024). The inclusion of predictive analytics and spatio-temporal forecasting has the potential to further empower the system to not only identify intrusions, but to more accurately predict potential threats through the analysis of vessel trajectories, behavioural anomalies and environmental data. Another important area of study is that of autonomous response mechanisms. By integrating detection capabilities with un-managed aerial vehicles (UAVs) or autonomous surface vessels (ASVs), faster detection and on-the-spot verification and response to intrusions is possible, which greatly reduces response times in high-stakes incidents (Nikolic et al., 2023). In addition, investigation of adaptive learning methods - where the system continually re-trains itself on new operational data - may lead to better robustness in coping with changes to threat behaviour, and mitigate malicious cyber manipulation. Finally, the creation and development of regulatory and ethical frameworks continues to be an important priority for research. Given that international guidelines for the deployment of AI surveillance in maritime zones are in an early phase, further work needs to be done in the areas of governance models, liability frameworks, and privacy-preserving algorithms that balance operational effectiveness with good, responsible technology use (Globisch et al., 2024).

12. CONCLUSION

The paper describes a Smart Intrusion Detection System (SIDS) to restricted maritime areas, which involves YOLO-based detection, optical flow, and computational modeling, with CNN-LSTM networks, to describe intrusions in space and time. Experimental analysis has shown that the system can be deployed on computationally limited edge nodes with promising accuracy and run time, thus highlighting that the system is appropriate to operational deployment. In addition to the technical discussion, the manuscript includes discussion of operational, cybersecurity and ethical considerations that fuel the adoption of AI-enabled surveillance systems. The authors represent SIDS as something that exceeds a technical artifact because it is more explicitly concerned with limitations pertaining to hardware, vulnerabilities raised in the process of data collection and transmission, and legal liability. Moving forward, the system could be implemented as a fully integrated multimodal maritime surveillance platform. It is expected that by deploying visual analytic, radar, satellite imagery, predictive models and autonomous response systems in synergy, leverage would be gained to provide unprecedented situational awareness. By integrating non-destructively these technological solutions combined with consideration of ethical, legal and governance issues, the proposed system provides the groundwork of a next generation intelligent, accountable and sustainable maritime surveillance ecosystem.

12.1. Summary of Findings

This study demonstrated the effectiveness of the motion tracking, spatio-temporal and deep learning methods on improving an intrusion detection in the restricted zone of the sea. In the evaluation of the proposed framework, it was shown to be more accurate, had a faster response time, and was more resilient to environmental and environmental-related challenges, when used on drone-based video feeds than the traditional surveillance methods. The implementation of a motion tracker using optical flow, and performer of anomaly classification using CNN LSTM decreased false-alarm rates, while demonstrating real-time computational speed. The findings indicate that intelligent AI-enabled surveillance can provide better security for ports and naval facilities and is a viable option to provide secure scalable solutions for offshore infrastructures (Prasad et al., 2017).

12.2. Final Thoughts on System Effectiveness

The usefulness of the proposed intrusion detection system is the ability to employ spatio-temporal analysis of motion and AI-based classification to address the challenges of operating in the maritime environments. The framework can be employed as an alternative to standard surveillance systems, unlike other situations, which are static in nature and do not adapt to changing sea conditions, visibility and the changing traffc of vessels. Addition of drones increases the span of monitoring, and quick response can be made in a wide limited area. Although computational limitations and the noise of the environment still pose a challenge, the system has great potential to be scalably deployed in naval bases, ports and offshore installations, and could serve as a realistic way of providing the next generation of maritime security systems (Colomina & Molina, 2014).

12.3. Implications for Maritime Policy and Safety

The use of maritime operations as a context of AI-based intrusion detection is also associated with essential policy and safety concerns. Edge intelligence provides on-device processing of sensitive sensor data, minimizing bandwidth overhead but transferring risks of cybersecurity, data governance, and attribution of incidents to operators and manufacturers and mandate interoperability. Regulators need to revise norms of drone operation, certification, cross border data sharing and rules of engagement in restricted zones. Auditability, resistance to adversarial attacks, and privacy protection of the coastal population, should be mandated using policy frameworks (Suresh Babu, C. V., Indhumathi, C. et. al. 2023).. Harmonization of technical guidelines and the creation of proper liability guidelines will fast-track safe implementation of AI-enhanced surveillance without compromising on maritime laws and human rights (Zhou et al., 2019).

REFERENCES

Ahmed, I., Jun, M., & Ding, Y. (2022). A spatio-temporal track association algorithm based on marine vessel automatic identification system data. *IEEE Transactions on Intelligent Transportation Systems*, *23*(11), 20783–20797. DOI: 10.1109/TITS.2022.3187714

Balduzzi, M., Pasta, A., & Wilhoit, K. (2014, December). A security evaluation of AIS automated identification system. In *Proceedings of the 30th annual computer security applications conference* (pp. 436-445). https://doi.org/DOI: 10.1145/2664243.2664257

Bloisi, D. D., Previtali, F., Pennisi, A., Nardi, D., & Fiorini, M. (2016). Enhancing automatic maritime surveillance systems with visual information. *IEEE Transactions on Intelligent Transportation Systems*, *18*(4), 824–833. DOI: 10.1109/TITS.2016.2591321

Bochkovskiy, A., Wang, C. Y., & Liao, H. Y. M. (2020). Yolov4: Optimal speed and accuracy of object detection. *arXiv preprint arXiv:2004.10934*. https://doi.org//arXiv.2004.10934DOI: 10.48550

Buczak, A. L., & Guven, E. (2015). A survey of data mining and machine learning methods for cyber security intrusion detection. *IEEE Communications Surveys and Tutorials*, *18*(2), 1153–1176. DOI: 10.1109/COMST.2015.2494502

Colomina, I., & Molina, P. (2014). Unmanned aerial systems for photogrammetry and remote sensing: A review. *ISPRS Journal of Photogrammetry and Remote Sensing*, *92*, 79–97. DOI: 10.1016/j.isprsjprs.2014.02.013

Globisch, T., Umlauft, A., Lüdtke, S., & Von Lukas, U. F. (2024, September). sMarT RCI-Unsupervised Double K-Means Ensemble Clustering Tool for Remotely Operated Vehicle Camera Images. In *OCEANS 2024-Halifax* (pp. 1-10). IEEE. https://doi.org/DOI: 10.1109/OCEANS55160.2024.10754492

Hall, D., & Llinas, J. (2001). *Multisensor data fusion*. CRC press., DOI: 10.1201/9781420038545

Kong, S. G., Heo, J., Abidi, B. R., Paik, J., & Abidi, M. A. (2005). Recent advances in visual and infrared face recognition—A review. *Computer Vision and Image Understanding*, *97*(1), 103–135. DOI: 10.1016/j.cviu.2004.04.001

Liang, M., Weng, L., Gao, R., Li, Y., & Du, L. (2024). Unsupervised maritime anomaly detection for intelligent situational awareness using AIS data. *Knowledge-Based Systems*, *284*, 111313. DOI: 10.1016/j.knosys.2023.111313

Liu, R. W., Liang, M., Nie, J., Lim, W. Y. B., Zhang, Y., & Guizani, M. (2022). Deep learning-powered vessel trajectory prediction for improving smart traffic services in maritime Internet of Things. *IEEE Transactions on Network Science and Engineering*, *9*(5), 3080–3094. DOI: 10.1109/TNSE.2022.3140529

Liu, R. W., Liang, M., Nie, J., Yuan, Y., Xiong, Z., Yu, H., & Guizani, N. (2022). STMGCN: Mobile edge computing-empowered vessel trajectory prediction using spatio-temporal multigraph convolutional network. *IEEE Transactions on Industrial Informatics*, *18*(11), 7977–7987. DOI: 10.1109/TII.2022.3165886

Mascaro, S., Korb, K.B., & Nicholson, A.E. (2010). Learning Abnormal Vessel Behaviour from AIS Data with Bayesian Networks at Two Time Scales.

Mascaro, S., Nicholso, A. E., & Korb, K. B. (2014). Anomaly detection in vessel tracks using Bayesian networks. *International Journal of Approximate Reasoning*, *55*(1), 84–98. DOI: 10.1016/j.ijar.2013.03.012

Nikolic, D., Stojkovic, N., Puzovic, S., Popovic, Z., Stojiljkovic, N., Grbic, N., & Orlic, V. D. (2023). Increasing Maritime Safety and Security in the Off-Shore Activities with HFSWRs as Primary Sensors for Risk Assessment. *Journal of Marine Science and Engineering*, *11*(6), 1167. https://www.mdpi.com/2077-1312/11/6/1167. DOI: 10.3390/jmse11061167

Prasad, D. K., Rajan, D., Rachmawati, L., Rajabally, E., & Quek, C. (2017). Video processing from electro-optical sensors for object detection and tracking in a maritime environment: A survey. *IEEE Transactions on Intelligent Transportation Systems*, *18*(8), 1993–2016. DOI: 10.1109/TITS.2016.2634580

Riveiro, M., Falkman, G., & Ziemke, T. (2008). Visual analytics for the detection of anomalous maritime behavior. In *Proceedings of the International Conference on Information Visualisation* (pp. 273–279). IEEE. https://doi.org/DOI: 10.1109/IV.2008.25

Riveiro, M., Pallotta, G., & Vespe, M. (2018). Maritime anomaly detection: A review. *Wiley Interdisciplinary Reviews. Data Mining and Knowledge Discovery*, *8*(5), e1266. DOI: 10.1002/widm.1266

Shi, W., Cao, J., Zhang, Q., Li, Y., & Xu, L. (2016). Edge computing: Vision and challenges. *IEEE Internet of Things Journal*, *3*(5), 637–646. DOI: 10.1109/JIOT.2016.2579198

Sommer, R., & Paxson, V. (2010). Outside the closed world: On using machine learning for network intrusion detection. *2010 IEEE Symposium on Security and Privacy*, 305–316. https://doi.org/DOI: 10.1109/SP.2010.25

Suresh Babu, C. V., Andrew Simon, P., & Barath Kumar, S. (2023). The Future of Cyber Security Starts Today, Not Tomorrow. In Shiva Darshan, S., Manoj Kumar, M., Prashanth, B., & Vishnu Srinivasa Murthy, Y. (Eds.), *Malware Analysis and Intrusion Detection in Cyber-Physical Systems* (pp. 348–375). IGI Global., DOI: 10.4018/978-1-6684-8666-5.ch016

Suresh Babu, C. V., Suruthi, G., & Indhumathi, C. (2023). Malware Forensics: An Application of Scientific Knowledge to Cyber Attacks. In Shiva Darshan, S., Manoj Kumar, M., Prashanth, B., & Vishnu Srinivasa Murthy, Y. (Eds.), *Malware Analysis and Intrusion Detection in Cyber-Physical Systems* (pp. 285–312). IGI Global., DOI: 10.4018/978-1-6684-8666-5.ch013

Suresh Babu, C. V., Suruthi, G., Indhumathi, C., & Sushruth, S. (2024). Integrated approach for network intrusion detection: Combining KNN algorithm, Karhunen-Loève transform, and genetic algorithm optimization. In Metaheuristic and machine learning optimization strategies for complex systems (p. 21). DOI: 10.4018/979-8-3693-7842-7.ch002

Szegedy, C., Vanhoucke, V., Ioffe, S., Shlens, J., & Wojna, Z. (2016). Rethinking the inception architecture for computer vision. In *Proceedings of the IEEE conference on computer vision and pattern recognition* (pp. 2818-2826). **https://doi.org/**DOI: 10.1109/CVPR.2016.308

Talpur, K., Hasan, R., Gocer, I., Ahmad, S., & Bhuiyan, Z. (2025). AI in Maritime Security: Applications, Challenges, Future Directions, and Key Data Sources. *Information (Basel)*, *16*(8), 658. DOI: 10.3390/info16080658

Tu, E., Zhang, G., Mao, S., Rachmawati, L., & Huang, G. (2020). Modeling Historical AIS Data For Vessel Path Prediction: A Comprehensive Treatment. *ArXiv*. https://arxiv.org/abs/2001.01592

Tu, E., Zhang, G., Rachmawati, L., Rajabally, E., & Huang, G. B. (2017). Exploiting AIS data for intelligent maritime navigation: A comprehensive survey from data to methodology. *IEEE Transactions on Intelligent Transportation Systems*, *19*(5), 1559–1582. DOI: 10.1109/TITS.2017.2724551

Venskus, J., Treigys, P., Bernatavičienė, J., Tamulevičius, G., & Medvedev, V. (2019). Real-time maritime traffic anomaly detection based on sensors and history data embedding. *Sensors (Basel)*, *19*(17), 3782. DOI: 10.3390/s19173782 PMID: 31480449

Wang, X., Ma, X., & Grimson, W. E. L. (2008). Unsupervised activity perception in crowded and complicated scenes using hierarchical bayesian models. *IEEE Transactions on Pattern Analysis and Machine Intelligence*, *31*(3), 539–555. DOI: 10.1109/TPAMI.2008.87 PMID: 19147880

Yao, H., Qin, R., & Chen, X. (2019). Unmanned aerial vehicle for remote sensing applications—A review. *Remote Sensing (Basel)*, *11*(12), 1443. DOI: 10.3390/rs11121443

Zhang, C., Patras, P., & Haddadi, H. (2019). Deep learning in mobile and wireless networking: A survey. *IEEE Communications Surveys and Tutorials*, *21*(3), 2224–2287. DOI: 10.1109/COMST.2019.2904897

Zhang, X., Liu, J., Chen, K., Gong, P., Liu, Y., & Wu, Z. (2024). Learning dynamic interactions and long-term patterns with spatio-temporal graphs for multi-vessel trajectory prediction. *IEEE Transactions on Intelligent Vehicles*, *9*(12), 7765–7780. Advance online publication. DOI: 10.1109/TIV.2024.3401864

Zhao, L., Song, Y., Zhang, C., Liu, Y., Wang, P., Lin, T., Deng, M., & Li, H. (2019). T-GCN: A temporal graph convolutional network for traffic prediction. *IEEE Transactions on Intelligent Transportation Systems*, *21*(9), 3848–3858. DOI: 10.1109/TITS.2019.2935152

Zhao, Z., Chen, W., Wu, X., Chen, P. C., & Liu, J. (2017). LSTM network: A deep learning approach for short-term traffic forecast. *IET Intelligent Transport Systems*, *11*(2), 68–75. DOI: 10.1049/iet-its.2016.0208

Zhou, Z., Chen, X., Li, E., Zeng, L., Luo, K., & Zhang, J. (2019). Edge intelligence: Paving the last mile of artificial intelligence with edge computing. *Proceedings of the IEEE*, *107*(8), 1738–1762. DOI: 10.1109/JPROC.2019.2918951

Zissis, D., Chatzikokolakis, K., Spiliopoulos, G., & Vodas, M. (2020). A distributed spatial method for modeling maritime routes. *IEEE Access : Practical Innovations, Open Solutions*, *8*, 1–1. DOI: 10.1109/ACCESS.2020.2979612

Zissis, D., & Lekkas, D. (2012). Addressing cloud computing security issues. *Future Generation Computer Systems*, *28*(3), 583–592. DOI: 10.1016/j.future.2010.12.006

Zissis, D., Xidias, E. K., & Lekkas, D. (2015). A cloud based architecture capable of perceiving and predicting multiple vessel behaviour. *Applied Soft Computing*, *35*, 652–661. DOI: 10.1016/j.asoc.2015.07.002

Zissis, D., Xidias, E. K., & Lekkas, D. (2016). Real-time vessel behavior prediction. *Evolving Systems*, *7*(1), 29–40. DOI: 10.1007/s12530-015-9133-5

Chapter 8
The Strategic Deployment of Drones in Taiwan's Maritime Domain:
Opportunities and Operational Benefits

Wes Harven Guillemer Maravilla
https://orcid.org/0000-0001-5592-3425
Suan Sunandha Rajabhat University, Thailand

ABSTRACT

This study investigates the strategic deployment of drones in Taiwan's maritime domain, analyzing their opportunities and operational benefits. The research explores the increasing integration of Unmanned Aerial Vehicles (UAVs) in the maritime sector and their impact on enhancing safety, efficiency, and productivity. Through a needs-based analysis and technology assessment, key application areas such as port monitoring, environmental surveillance, and search and rescue operations are identified. The study also addresses the challenges and barriers to drone implementation, including regulatory considerations, technological limitations, and social-economic factors, proposing solutions and mitigation strategies to facilitate effective drone integration in Taiwan's maritime sector.

INTRODUCTION

Changes so far So called unmanned aerial vehicles (UAV) – or drones – are becoming more and more widespread, and the "maritime industry is poised to reap the benefits of its use on a huge scale". In addition to drone rescues, droid trans-

DOI: 10.4018/979-8-3373-3546-9.ch008

portation is becoming increasingly popular. In order for the maritime industry to operate efficiently and effectively, they must be integrated.

Considering where Taiwan's marine industry is located, it is impossible not to use UAVs in operations. There has been an increase in smuggling and congestion in the larger ports like Kaohsiung, Taipei, and Taichung. Drones are another force multiplier for Taiwan, because they can efficiently detect and track threats and provide continuous overhead surveillance.

There are a number of ecological pressures facing Taiwan's coasts and seas, including pollution and coastal erosion, as well as security concerns and logistical issues. The sensors on aircraft that carry these sensors allow us to chase these atmospheric conditions, giving us a technological edge. It is important for the environment and for preserving our coastline. The coast is also a hot spot for SAR (search and rescue) missions. In addition to arriving faster in an emergency, drones are also equipped with life-saving equipment that makes Taiwan more resilient.

Drones are going to become more important in the next few years, as we use them more and more (Verified Market Reports, 2025).

DRONE APPLICATIONS IN TAIWAN'S MARITIME OPERATIONS

With the rise of unmanned aerial vehicles, or drones, the maritime industry could experience a fantastic transformation in the near future. The use of drones for emergency response and transportation is becoming more commonplace - and more advanced. For maritime activities to be more efficient, effective, and safe, these systems need to be interoperable.

Taiwan's marine industry relies heavily on UAVs due to its geographic location. Historically, large ports like Kaohsiung, Taipei, and Taichung have experience congestion and smuggling issues due to their high cargo traffic. In addition to drones, Taiwan has another force multiplier in the form of surveillance drones, which are more effective at scanning and tracking threats.

In addition to security and logistical challenges, Taiwan's coasts and seas face unique ecological pressures, such as pollution and coastal erosion. Our airplanes are equipped with very specialized sensors, so we can monitor these environmental drivers. This is an imperative step to protect our environment and preserve our coast. SAR (search and rescue) operations are also possible along the coast. Taiwan can better respond to emergencies when drones carry life-saving equipment as well as shorten response times. In the coming years, drone use will become more comfortable, making these features increasingly important (Verified Market Reports, 2025).

BENEFITS AND CHALLENGES OF DRONE INTEGRATION IN TAIWAN'S MARITIME SECTOR

Drones have started to make a real difference during rescue missions—they can find people fast and even drop emergency gear right where it's needed (G S K et al., 2024). Maritime SAR operations from the coast are also possible. Life-saving equipment on multi-copters can make Taiwan much more responsive anytime there is an emergency. They can also shorten the time it takes for Taiwan to respond. By 2020, drone technology will become easier to use, which will make these flight features more relevant (Verified Market Reports, 2025).

UAVs and drones are revolutionizing the maritime industry. To achieve a safe, efficient, and productive system, it is important to collaborate between these two technologies. With Taiwan's seascape as a backdrop, drones have the potential to be deployed strategically in both positive and negative ways. Another benefit of using drones for property inspection is that it can save time and money, as well as make the process safer. Furthermore, Taiwan reduces environmental impacts in order to achieve its sustainability goals. In spite of this, Taiwan's maritime sector faces a number of stumbling blocks. The combination of drones with communications (reliable and robust) equals red (highly regulated) in addition to being grounded in Law. In order for integration to be successful and responsible, security, privacy, and technical concerns also have to be addressed.

Table 1. SWOT Analysis of Drone Integration in Taiwan's Maritime Sector

	Helpful	**Harmful**
Internal Origin	**Strengths (S)** ***Enhanced Safety** ***Operational Efficiency and Cost Savings** ***Versatile Applications** ***Advanced Data Collection**	**Weaknesses (W)** ***Technological Limitations** ***Maneuverability and Human Oversight** ***Communication Challenges**
External Origin	**Opportunities (O)** *Economic Growth *Leveraging Emerging Technologies *Policy and Collaboration	**Threats (T)** *Regulatory Ambiguity *Privacy and Security Concerns *Initial Investment and Workforce Skills

Also analyzed Taiwan's naval drones strategy according to SWOT (strengths, weaknesses, opportunities and threats). It also identifies drone technology's and processes' best and worst points, but this includes both external and internal strengths. In contrast to a simple list of applications and benefits, critical and strategic considerations give insight into why, based on which enabler, future drone integration will take place. These days, drones aren't just flying around for fun—they're actually making it way easier to move stuff between ships and shore, check out boats without risking workers, keep an eye on the environment, and jump into action fast when people need help at sea. Plus, they're saving time and cutting a lot of costs for everyone (IJTSRD, 2024).

Strengths

- Enhanced safety: Drones can also inspect tight quarters or hazardous environments, like ships' hulls and cargo tanks, which reduces the risk of humans having accidents during similar assessments. Taiwan's shipyards need it especially. Additionally, drones can be used in SAR to decrease response time and find people.
- A drone inspection can be completed 90-95% faster than conventional methods, saving you time and money. In addition to saving money, this efficiency also means less downtime for ship owners and the shipping industry at large.
- You can use drones for everything from monitoring ports to inspecting cargo to transporting supplies. Their uses can vary based on their context, whether they're semi-autonomous or completely autonomous.
- Drones can collect high-quality data in real time with advanced cameras and sensors (LiDAR, for example). Inspecting, monitoring, and surveillance are all important.

Weaknesses

- Because drone batteries run out, they have a limited run time. Drones can only be used for a few minutes before they have to be recharged. Taiwan's marine environment struggles because of bad weather. For more robust gulpers, they're too flimsy.
- Despite drones' agility and human oversight, some tasks can't be automated, so a pilot is a must. Some places are also hard to find.
- Communication problem: It's tough to get the same control and data transfer performance over vast oceans or congested areas. Validated protocols must be used to protect drone data and their communications.

Opportunities

- Growing industry: Drones are booming. Through drone production, operations, and data analysis, Taiwan can save money and create jobs.

- Every aspect of drone technology is getting better, from navigation to flight control. Using artificial intelligence and 5G, drones could be fully autonomous and communicate in real-time.
- To build an integrated regulatory system, Taiwan needs to collaborate (and collude) with organizations like EMSA. To speed up the process of entering new technologies, many tentative solutions consider the roles of government, industry, and academia.

Threats

- Drones couldn't be imported into Taiwan because of regulatory uncertainty. It makes investment and market take-up more expensive because of this uncertainty.
- Drone privacy and security issues exist: Drones have implications for privacy and security. In order to protect drone data from unauthorized access, it needs to be secured.
- Investing in new skills/budget is a major barrier for some stakeholders. Training programs are needed for these systems beyond their operators and maintainers.

Taiwan also has a marine industry that can benefit from the use of drones. According to the SWOT analysis, drone technology has many advantages, including the ability to collect high-level data and prevent risk. Although Taiwan must be proactive and multifaceted to take advantage of these opportunities and compensate for external threats such as ambiguous regulations. Taiwan can mitigate the identified challenges by investing in technology, establishing clear regulatory frameworks, and fostering public-private collaboration, putting it at the forefront of maritime innovation, ensuring a more secure and resilient maritime future.

TECHNOLOGICAL ASPECTS OF DRONES IN TAIWAN'S MARITIME USE

These days, drones are getting really smart at spotting things on the water all by themselves, which is making everything in Taiwan's marine world run a lot smoother and safer (Panico, Fragonara, & Al-Rubaye, 2020). Drone technology is key to Taiwan's effective integration of drones in maritime operations. A drone can be semi- or fully autonomous, and there are different types, like multirotor drones, fixed-wing drones, single-rotor drones, and fixed-wing hybrid VTOLs, each with their own advantages and disadvantages. Developments in drone technology have led to advancements in flight control technology, navigation, power technology, communication systems, and vision systems.

To ensure a strong, clear, and high-impact investigation, the research objectives have been refined to be more specific and measurable, directly aligning with the study's methodology and findings.

- To identify and prioritize key areas within Taiwan's maritime sector where drone technology can be strategically deployed to enhance operational efficiency and security, based on a needs-based analysis.
- To evaluate the specific operational advantages of integrating drones, including quantifiable improvements in safety and efficiency, by conducting a comprehensive technology assessment.
- Develop data-driven solutions and mitigation strategies based on comparative case studies and best practices, addressing the major challenges and barriers to drone implementation in Taiwan.

DRONE APPLICATIONS IN MARITIME OPERATIONS

In the maritime sector, drones have transformed traditional practices and offered innovative solutions. UAVs can be used for routine inspections, such as hull surveys, cargo and water ballast tanks, and structural integrity, enhancing safety and efficiency. It's common for drones to be used for these inspections because they're equipped with high-resolution cameras and non-destructive testing (NDT) sensors. Their services include finding lost individuals, rescuing men overboard, delivering life jackets or life rafts, and improving overall response times. These new types of sensors on drones are a game changer—they help crews spot ships and other stuff in real time, even when the weather's rough or it's super dark out. It's made things way easier for folks working out on the water in Taiwan (Bai, Zhang, & Xu, 2025).

Additionally, drones are capable of tracking traffic, smelling pollution, and delivering goods, papers, spare parts, and emergency medicine to ships at sea and in harbors. The maritime industry can benefit from drones in so many ways. An inspector can use a multirotor or fixed-wing drone to survey difficult-to-reach areas and relay the images in real time to the inspector. Using LiDAR (Light Detection and Ranging), 3D surfaces can be mapped precisely, whereas thermal imaging cameras can be used at night or in low light.

Furthermore, drones can be used to control border coastal/offshore immigration, monitor harbours (aerial videography for loading operations, thermographic cameras for detecting people), and detect incidents over the open ocean, including illegal fishing and drug trafficking, demonstrating how they can improve security and surveillance. These operations often rely on secure communication protocols to ensure data integrity and prevent unauthorized access to the drone or its data . Overall, the integration of drones offers significant potential for optimizing maritime operations, improving safety, and promoting environmental sustainability.

Table 2. Summary of Drone Applications in the Maritime Sector

Application Area	Description of Drone Use	Benefits	Challenges
Inspections	Routine assessments of ship hulls, cargo holds, and infrastructure using high-resolution cameras and sensors.	Enhanced safety and efficiency in inspections, reduced downtime.	Potential need for skilled operators, data management, and analysis complexity.
Search and Rescue	Finding missing persons, assisting in distress situations, and delivering life-saving equipment.	Improved response times and coverage, increased effectiveness.	Difficulties in extreme weather conditions, limited endurance for long-range operations.
Port Operations	Monitoring traffic, managing logistics, detecting spills, and ensuring security.	Support for efficient port management and safety, and real-time situational awareness.	Ensuring reliable communication in congested environments, integrating with existing port infrastructure.
Open Sea Operations	Delivering supplies, monitoring environmental changes, and supporting research activities.	Support for remote operations and access to difficult-to-reach areas.	Limited operation time due to battery power and payload capacity restrictions.
Specialized Tasks	Conducting surveillance, enforcing regulations, and assessing damage after incidents.	Enhanced security and surveillance, improved environmental monitoring, and law enforcement.	Regulatory and legal uncertainties, concerns regarding privacy and data security.

Table 2 effectively illustrates the multifaceted role of drone technology within the maritime sector, highlighting its capacity to redefine traditional operational boundaries. The table goes beyond a simple listing of tasks; it infers a shift towards a more data-driven, efficient, and responsive maritime industry.

However, it also subtly emphasizes that the integration of drones is a complex undertaking. It is not just about introducing new tools but about navigating a web of operational, technological, and regulatory changes. Drones' potential can only be realized by embracing their capabilities and proactively tackling their deployment's inherent challenges.

EAST ASIAN DRONE POLICY AND REAL-WORLD MARITIME INCIDENTS

It is becoming more common for East Asian nations to develop policies and regulations for managing drone deployments in the maritime sector. With advanced 4G and 5G networks, Singapore is becoming a global leader in maritime drone innovation. Collaboration between government, industry, and technology providers is vital to fostering drone adoption and ensuring real-time data transfers (Maritime and Port Authority of Singapore, 2020). It is clear that clear regulations are required to support advanced drone applications in port logistics, as demonstrated by Taiwan's initiatives.

There is a need for clear drone policies after drones were used in high-stakes maritime incidents. The use of drones in SAR operations has improved the speed with which people have been located and emergency supplies have been delivered following maritime accidents by locating missing people quicker than traditional methods. In Australia, two teenage swimmers caught in rip currents had flotation devices delivered to them by drones. Additionally, drones are used to detect illegal activities such as fishing and smuggling through maritime surveillance and enforcement. Zhejiang, China, police used VTOL drones to monitor illegal fishing during a four-month fishing ban, covering vast areas of water and reporting it in real time to the ground control center. As a result of these real-world drone applications, Taiwan and the broader East Asian region can develop comprehensive drone policies that enhance maritime security and safety.

TECHNOLOGICAL ASPECTS OF DRONES IN MARITIME USE

Technology plays a big role in drone integration into maritime operations. There are several types of drones out there, including multirotor drones, fixed-wing drones, single-rotor drones, and fixed-wing hybrid VTOLs, all with their own characteristics, advantages, and disadvantages (Krystosik-Gromadzińska, 2021). It's possible to do a lot in the maritime sector with the right drone and equipment.

There have been a lot of advancements in drone technology. Among the things they're working on are flight control, navigation, power, communication and vision systems. Citespace's first phase measures keyword frequency. To maximize drone's potential in maritime, we have to address the technical challenges and leverage the advances in drone technology.

Table 3. Comparative Analysis of Drone Types

Drone Type	Advantages	Disadvantages/Limitations	Typical Maritime Applications
Multirotor Drones	Maneuverability, vertical take-off and landing (VTOL), and hovering capability.	Limited flight time compared to fixed-wing, lower speed.	Inspections, surveillance, search, and rescue in localized areas.
Fixed-Wing Drones	Longer flight time, higher speed, greater range.	Requires a runway or launch/ recovery system, less maneuverable than multirotors.	Long-range surveillance, coastal monitoring, and delivery over longer distances.
Fixed-Wing Hybrid VTOLs	Combines the advantages of both fixed-wing (long flight time, speed) and multirotor (VTOL, hovering).	Complexity in design and control generally results in a higher cost.	Versatile applications requiring both range and maneuverability, such as offshore inspections and delivery to ships.

An overview of the major drone categories relevant to maritime applications is shown in Table 3, which highlights the trade-offs between their design and functionality. Types of drones are selected based on the specific requirements of the maritime task at hand, not arbitrarily. Multirotor drones, for example, excel in situations that require precise positioning and close-range observation, while fixed-wing drones work better for expansive surveillance or rapid delivery. By combining the vertical capabilities of multirotors with the efficiency of fixed-wing aircraft, multirotor hybrids provide operational flexibility. Drone designs demonstrate UAV technology's versatility in meeting multifaceted maritime challenges.

REGULATORY AND LEGAL CONSIDERATIONS FOR DRONE OPERATIONS AT SEA

The use of drones in Taiwan's maritime industry is being held back by substantial challenges related to regulation and law. The absence of transparent and holistic regulation has dumbed down the operators' community in the country leading to deprivation of incentivisation to invest in DRONE as an upcoming technology deployment capabilities. Regulatory agencies should also address the grey areas of operations, liability and data protection within the legal environment to be conducive for wide use of drones over waters.

Overcoming such barriers requires construction of a flexible framework that adequately meets the balance of technological innovation and public safety. It is highly desirable that Taiwan adopt those international best practices as outlined by the EMSA, particularly on operational areas, flight permissions and liability principles. "Incorporating those international standards, Taiwan can offer the sufficient certainties for the drone operators to be introduced in our maritime domain safely and responsibly."

METHODOLOGY

Following is a narrative review flow chart adapted from PRISMA that illustrates how the process is done. Throughout it, you'll find an overview of the process of identifying, screening, determining eligibility, and synthesizing records.

Figure 1. PRISMA-style flowchart of a literature review.

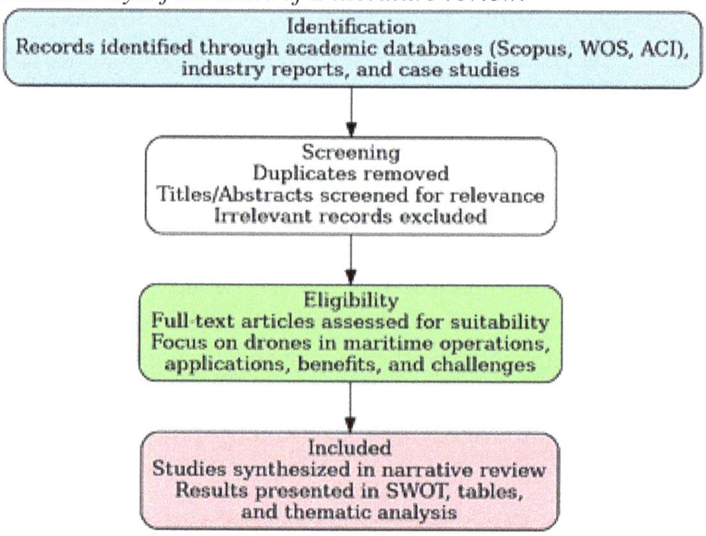

Our analysis is based on a narrative review of the existing literature and research on drones, specifically those deployed in the maritime sector in Taiwan. To give a comprehensive and holistic view of current applications, benefits, and challenges, we integrate findings from academic studies, industry reports, and case studies. Statistical meta-analyses are not appropriate for nascent fields like maritime drone applications. Instead, narrative reviews provide a broader synthesis of themes.

We searched and analyzed relevant literature to identify key themes and patterns that directly addressed the research questions. Our study uses this method to offer insights into the operational benefits and opportunities of drone technology in Taiwan's maritime sector for policymakers, industry stakeholders, and researchers.

RESULTS AND DISCUSSION

Identifying Drone Applications in Taiwan's Maritime Sector: Results of Needs-Based Analysis

Figure 2. Prioritized Drone Applications in Taiwan's Maritime Sector

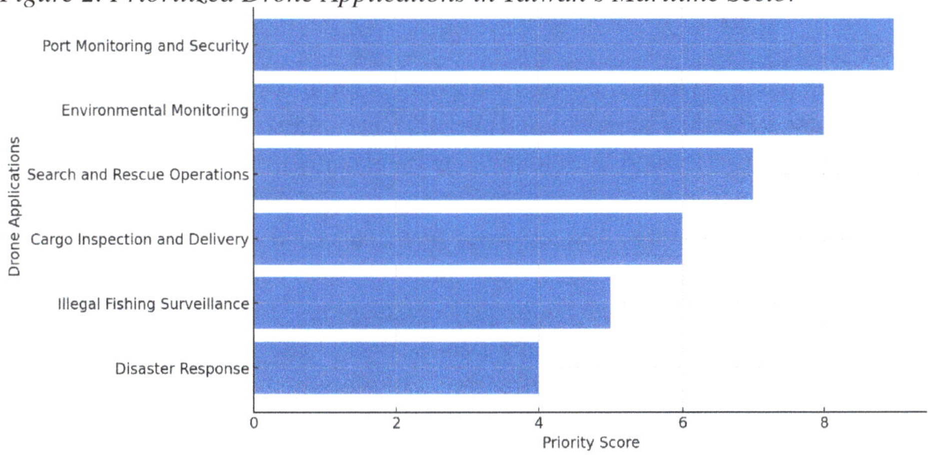

Taiwan's maritime sector analysis shows that drone applications are prioritized based on the nation's most pressing security and operational needs. As far as drone deployment goes, port monitoring, security, and environmental monitoring are the top priorities. In the Figure above, you can see how these applications protect coastal regions from pollution and safeguard critical infrastructure. Safety and efficiency are also improved through Search and Rescue Operations and Cargo Inspection and Delivery.

Monitoring and security of ports were improved with increased perimeter surveillance, coordinated traffic flow management, and scrutiny of ships. This introduction will benefit Taiwan's main ports like Kaohsiung, Taipei, and Taichung, as well as locating drones and their controllers through GPS by tracking flight paths. For Taiwan's maritime ports to remain operationally viable and safe, these capabilities are critical.

In addition to environmental observation, drones can also be used for coastal management. It is possible for drones to detect sulphur and carbon dioxide pollution in Taiwan's waters! Using drones, satellite imagery, and unmanned underwater vehicles, we will be able to gather environmental data, map the coast, and survey marine habitats.

Throughout Taiwan's coast, drones could also assist search and rescue operations. In the early stages of SAR, drones have already been used to search for missing persons, man overboard cases, and post survival kits. Taiwan's experience with drones can be generalized to show how they may improve security, safety, and efficiency in maritime activities.

Technology Assessment of Drones and Operational Advantages

Taiwan's maritime market is heavily dependent on the use of drones, both from an efficiency and safety standpoint. Taiwan's shipbuilding industry — based in Kaohsiung — uses a lot of drones to mitigate the risk involved with human inspections. Drones with high-definition cameras and sensors can scan ship structures, including cargo and water ballast tanks, for rust, cracks and other damage allowing workers to avoid risking their lives. That added capability means that Taiwan can turn to drones for things that are “dull, dirty or dangerous” for humans to do -- with safer working conditions and fewer accidents.

Taiwan's maritime market benefits from drone use both from a safety and efficiency standpoint. To reduce risks associated with human inspections, Taiwan's shipbuilding industry, centered in Kaohsiung, deploys clusters of drones. Using high-definition cameras on drones equipped with sensors, workers can inspect ships' structures, including cargo and ballast water tanks, for rust, cracks, and other damage without risking their lives. The added capability allows Taiwan, for instance, to use drones to do “dusty, dirty, and dangerous” work - with safer working conditions.

A drone also saves time and money, as demonstrated by the evaluation. Despite their enhanced inspection capabilities, drones may still necessitate the presence of a professional pilot and live surveyor, and their maneuverability can be challenging in certain situations, as noted by Smogeli (DNV GL, as cited in Vessel Performance Optimisation, 2020). There's still room for drone technology to be developed and refined. However, drones are driving innovation and competitiveness in Taiwan's dynamic maritime domain due to their enhanced safety, time, and cost savings.

Figure 3. Drone Technology Adoption Benefits in Taiwan

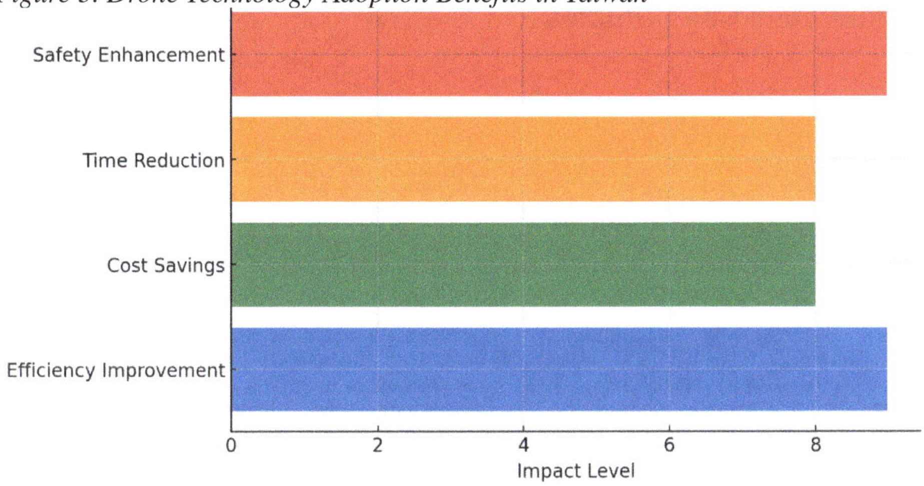

Taiwan's maritime sector has benefited from drone technology across a wide variety of operational areas. It's clear from the technology assessment that drone adoption will improve safety and efficiency. Drones optimize maritime logistics and security, streamline complex processes, and minimize human intervention in hazardous areas by leveraging automation and real-time data collection.

It's easy to see why this is the case in the Figure above, which shows how drone adoption has benefited Taiwan's maritime sector the most. In addition to immediate improvements, drones reduce resource waste and improve regulatory compliance in the long run. A rapid assessment and precise insight ensure better decision-making, which ultimately strengthens Taiwan's maritime infrastructure. Drone technology is expected to bring more efficiency and resilience to the sector as it evolves.

Best Practices in Maritime Drone Implementation: Comparative Case Studies

Table 4. Comparative Analysis of Maritime Drone Implementation

Feature	European Maritime Safety Agency (EMSA)	Singapore (MPA)	DNV-GL (Classification Society)
Primary Use Case	Maritime surveillance, monitoring pollution, and detecting illegal activities over vast areas.	Port operations, logistics, and shore-to-ship deliveries.	Ship inspections and structural integrity surveys.
Key Technology Focus	Drones are integrated into existing surveillance frameworks.	Use of advanced communication networks (4G/5G) for enhanced data transfer and real-time operations.	Specialized drones (e.g., Flyability Elios) and custom-built drones for confined spaces.
Operational Scope	Broad, regional surveillance over open seas (e.g., English Channel, North Sea, Baltic Sea).	Focused on a high-traffic port environment and specific trial use cases.	Global application for inspections of various vessel types, including tankers and bulkers.
Strategic Approach	Improving efficiency and enforcement capabilities by augmenting existing surveillance with drones.	Fostering a collaborative ecosystem between government, industry, and technology providers to test innovative use cases.	Developing standardized procedures and providing training for safe and effective implementation of inspections.

A table called "Comparative Analysis of Maritime Drone Implementation" provides an overview of three different approaches to drone integration, instead of presenting new findings. Its main purpose is to compare different strategies side-by-side and see what works. By using this visual format, it's easier to identify how major maritime organizations like EMSA, Singapore, and DNV-GL have implemented drones. What's even cooler is that some places are experimenting with using teams of drones to work together during tough rescue jobs, which can really boost how fast help arrives (Sneha & Surekha, 2025)

The table shows that while all three entities use drones, they approach it differently. Singapore, for example, focuses on technology infrastructure and port logistics instead of EMSA's wide-area surveillance and enforcement. It also emphasizes DNV-GL's global, application-specific approach to ship inspections, underscoring the importance of standardized procedures and training. This organized presentation of data helps a reader quickly grasp the varying priorities and strategic decisions behind each case. The table thus serves as a critical tool for synthesizing complex information into a clear and actionable format, which is essential for informed decision-making and for guiding future research directions.

Lessons for Taiwan

The experiences of these leading organizations provide valuable lessons for Taiwan's strategic deployment of drones in its maritime sector:

- **Integrate Drones into Existing Frameworks**: Similar to EMSA, Taiwan should integrate drones into its existing maritime surveillance systems to enhance domain awareness over its busy shipping lanes and coastal regions, improving efficiency and enforcement capabilities against illegal activities.
- **Make investments in technological infrastructure and collaboration:** Taiwan could follow Singapore's example by building 5G networks in its major ports to support advanced drone applications. Collaboration between government, industry, and universities is necessary to foster innovation.
- **The government of Taiwan should develop standardized procedures for drone use and offer specialized training to drone operators, especially for ship inspections, in order to ensure that drones are safe and effective.**

Breaking Into the Headwinds: The Problems of "smooth" Drone Practice in Taiwanese Maritime Shipment

The rough road ahead to an unmanned future on the seven seas in Taiwan From technical hurdles to logistical headaches, there are stormy skies in the offing for a drone-boats future of freedom for the Taiwan maritime sector. Operationally, the main obstacle is drones' finite charge that limits how far and how long they can fly. Furthermore, their performance is affected by inclement weather on the sea of Taiwan. A reliable communication system between the Daughters and the operator is still an open matter, especially in open-sea or busy ports. Some of those high level tasks are still going to be done with a human in the loop (professional pilot and live surveyor is recommended, for example) and there's no denying that flying a drone can be fiddly in some operating environments.

Besides, travel and scaling issues, the move to further drone use is restricted by a number of external and internal policy and cost barriers. A major external threat is the uncertainty of the regulatory environment: Unclear rules in general for using drones in Taiwan and heavy regulations have led to a lot of customer-driven hesitancy, thus no broad investment or purchase. This is compounded by the privacy and security issues arising from the drone data that must be protected against unauthorized access. The technology cost of value addition is what some entrepreneurs and stakeholder feel to be a lot inside the entities; much right there in the workforce is a demand for those skills, and its person-power handling programs that train operators/ servicers.

There are many technical, legal and financial impediments Taiwan must overcome to maximize the use of space technology in digital space projects.

Addressing the Challenges: Proposed Solutions and Mitigation Strategies

It's also important to have a unified management approach if Taiwan wants to solve problems with drones. In order for drones to operate effectively, we must have clear, strong regulations and legal standards; this will require cooperation among government departments, industry participants, and lawyers. Taiwan can embrace innovation while prioritizing safety, security, and privacy by following other jurisdictions' lead, such as the EU's Unmanned Aircraft Systems framework. It would likely be necessary to establish dedicated airspace, grant flight permissions, and establish liability rules to ensure that drone operators know their responsibilities. Investing in R&D and developing partnerships is crucial for Taiwan to overcome roadblocks such as short battery life, bad weather performance, and low carrying capacity. A robust and secure communication system - perhaps 5G networks or satellite communications, for example - must also be in place so that drone control and data can easily be exchanged in even the most remote sea areas.

Economic and Social Impact Assessment of Drone Deployment in Taiwan

Figure 4. Potential Economic and Social Impacts of Drone Deployment in Taiwan

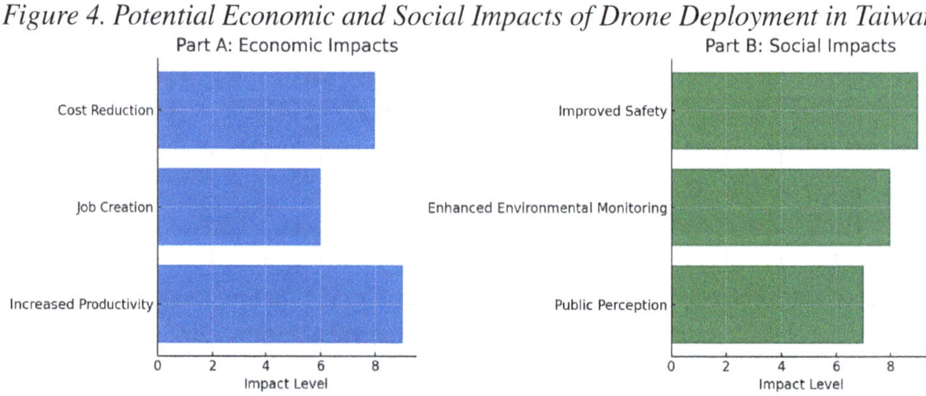

UAVs can be a great asset for Taiwan's waters. From an economic standpoint, the technology reduces costs and improves quality by automating jobs like inspection and monitoring. It'll take trained workers to manufacture, maintain, operate, and analyze drone data. In terms of society, SAR (search and rescue) and surveillance by drones will contribute to saving the lives of humans and the protection of property. In addition to offering real-time data on pollutants and marine conservation projects, they're crucial for environmental monitoring. However, in order to benefit from these positive effects, Taiwan needs investments in R&D, infrastructure and specialized skills. The tech for these sea drones just keeps on getting better, with smarter features popping up all the time, and it's actually giving Taiwan's whole maritime industry a big boost (Reports Insights Consulting Pvt Ltd, 2025).

Synthesizing Findings and Addressing Research Questions

Drones are a possibility, but there's a lot of resistance between now and when Amazon's self-driving drones can board ships. Drones aren't just about powerful, stable communications at sea. They're about a bunch of other things too. A lack of guidelines makes investing in and deploying drones on a large scale difficult in Taiwan. The concerns about privacy, data security, and initial investment costs must also be addressed and reduced in order for drones to be integrated safely and responsibly.

Regulatory frameworks need to be flexible and all-encompassing in order to deal with these challenges. In order to develop this framework, global best practices and rules, such as those developed by the European Union for unmanned aircraft systems, should be used. It is essential that drone operators have some kind of framework that provides them with some degree of certainty so that they can serve national interests.

Policy Implications for Taiwan

It is important for Taiwan to make short-term policy adjustments in order to make drones more accessible, particularly by enacting robust and flexible regulations. Right now, there's a lot of uncertainty around permissions, liability, and data security. Taking lessons from international bodies like the European Union guidelines on unmanned aircraft systems can help design clear regulations that strike a balance between innovation and safety. In addition to that, Taiwan needs to work on resolving conflicts of interest between the port authorities/ commission, unions and private operators; redefine roles and responsibilities to make sure labor rights are maintained, including safety standards; and make sure new regulations make it clear what's who's who. Additionally, test flying experiments and public education influence how the public adapts to drones, such as keeping them safe, protecting privacy, and reminding people of the opportunities for our maritime economies.

What will Taiwan's future look like? It's time to invest in its people and technology beyond the cross-party squabbles. The government has to "increase their funding and research in order to overcome barriers like battery life, weather resistance, and lifting capacity," he said. Also, drones could get more resilient and have longer-lasting power systems. In order to guarantee data integrity, it's also important to develop secure and trusted communication systems based on 5G or satellite communications. Policy makers will need to continue investing in specialized training programs and educational curricula to train seafaring professionals who will operate, maintain and process data from maritime drones such as those mentioned above – significantly increasing our rate of adoption. Taiwan's not messing around with drone tech lately—there's a big push from the government and other groups to use these things for keeping watch on the borders, responding super quickly to problems, collecting info, and showing they're serious about security across the strait (Brar, 2025).

Short-Term Policy Actions

Establish a Comprehensive and Adaptable Regulatory Framework. The most urgent need Taiwan has is clear rules for the operation of drones. Currently, drone usage is limited by murky regulations that prevent investment and usage. The government should also establish clear rules for crisscrossing flight paths, permissions for operations, and oversight of data security and liability. Unmanned aircraft systems standards developed by the European Union, for example, could provide a model for a framework that strikes a balance between innovation and safety, as well as national security.

Resolve the steelholders' conflict. Drones are becoming more popular, and that brings politics into play. Various concerns about job displacement and operational control have yet to be addressed, with policymakers ideally serving as facilitators in discussions between port authorities, unions, and private operators. For seamless fusion to take place, a new law and associated regulations will be necessary, as well as defining the roles and responsibilities of key stakeholders.

Conduct public awareness campaigns and pilot programs. Government should conduct pilot projects in critical maritime sectors and demonstrate drones' benefits. Through public information programs, citizens are educated on the merits of drone technology, as well as issues of privacy and safety.

Long-Term Strategic Investments

Drone consolidation also requires a long-term vision that includes strategic investment in technology and human capital.

Research: Will Taiwan invest in research and development to address its currently debilitating limitations (battery life, ability to withstand harsh weather, how much weight they can carry). We need policy that enables long-duration power systems and more robust drones against weather adversities. To ensure the integrity of data throughout oceans, it is also crucial to promote the use of safe and reliable communication systems, such as harnessing new technologies like 5G or satellites.

That's a big deal in terms of human capital development." It's important to have a skilled workforce that can operate, maintain, and analyze data from maritime drones. It's important to invest in specialized training programs and curriculums to give maritime professionals the skills they need.

Future Research Directions

It's a good idea to develop and validate interoperable regulatory frameworks and international collaboration models for cross-border maritime drone operations. Drones are becoming more common, so harmonized regulations will be necessary to ensure seamless and legal operations (Pastra et al., 2025). Comparing national and international drone regulations to find best practices for mutually recognising pilot certifications and operational permits could be part of this research.

In addition, developing AI-driven autonomous decision-making systems for maritime drones is an important future research area. In current applications, drones are mostly controlled remotely or semi-autonomous, but in the future, fully autonomous drone swarms could collaborate on large-scale surveillance, search and rescue operations, or environmental monitoring tasks without constant human supervision. We'd need to research more sophisticated algorithms to coordinate multi-agent missions, real-time adaptive mission planning, and fault-tolerance mechanisms. The long-term socioeconomic and environmental impacts of drone deployment should also be explored. The study discusses benefits like cost reduction and safety, but future research could measure these impacts over extended periods more precisely. Research could include examining the relationship between drone surveillance and illegal fishing reductions, or assessing drone manufacturing, deployment, and disposal's environmental impact.

Also, future research should look at how pervasive drone deployment affects the environment and socioeconomics. Future research could quantify these benefits over a longer period of time to quantify cost reductions and improved safety more precisely . Check out drone surveillance, illegal fishing, and drone manufacturing, deployment, and disposal for some ideas.

CONCLUSION

This is where we demonstrate that drones are a game-changer for for Taiwan's maritime industry; they offer quantum leaps in safety, productivity and environmental stewardship. We know these systems are doing a lot more than simply watching, efficiently overseeing such things as port security and complex cargo inspections to critical search-and-rescue missions. If it strategically invests in UAV deployment, Taiwan not only can ensure its key coastal areas becomes part of a sustainable future with real-time environmental data on conservation. Finally, the research is clear: drones fit neatly into Taiwan's fundamental strategic vision of building a modern and resilient maritime economy.

But tap this potential and Taiwan must first confront the immediate obstacles. "We' ve set a really ambitious but achievable target by 2030 because we want to make sure people have time to prepare for it," Ms Phillips said, adding the government wanted to give "business and industry more certainty" about decarbonising. The evidence is overwhelming that Taiwan now stands at a crossroads: through decisive, broad-based action — most crucially by setting clear rules and making intelligent technology choices — the country can transform itself into a global leader in ocean innovation.

Recommendation

Singapore has subscribed to this line of thinking, and Taipei should follow that approach immediately with targeted pilot programs in key maritime areas — such as in major ports and on susceptible coastlines. This includes proving such a value with drones such as performing ultrahigh-performance ship inspections and enhancing real-time port traffic control. In addition, there is a demand for the government to set up skills development programs/teaching curricula or other educational program that allows people to learn and prepare with new capability in operating and maintaining these advance systems. This mix of hands-on trials and workforce development is the fastest way to work the technology into daily operations.

The most immediate measure is the introduction of a coherent, but light touch, regulatory framework for drone use. That means establishing clear rules about everything from flight paths and operational permissions to who pays (liability) when something goes wrong. To quickly formulate this framework, Taiwan should refer to the existing international standard established, for example, by the European Union. Well thought-out guidelines are important both to take away the uncertainty that attracts investment and to ensure public concern about data security and privacy is addressed in a formal manner protected by law.

Future studies need not to be mere proofs of the concept, but should concentrate on next-steps challenges. Exploration of AI enabled autonomous drone swarms that can collaboratively plan large scale operations with only occasional human input, to be performed by academics. Another key area is the need for longer-term socioeconomic impact studies that allow a clear comparison of how much drones contribute to job creation, cost savings, and compliance with regulations over the long haul, giving policy makers measurable evidence of their intrinsic value.

REFERENCES

Bai, X., Zhang, Z., & Xu, X. (2025). Advanced multispectral detection of maritime targets for unmanned ocean surveillance. Ocean Engineering. https://www.sciencedirect.com/science/article/abs/pii/S0029801825008984

Brar, P. (2025). Drones Over the Strait: How Taiwan's UAV Programme is Redrawing Cross-Strait Red Lines. RSIS Commentary. https://www.rsis.edu.sg/wp-content/uploads/2025/08/CO25171.pdf

International Journal of Trend in Scientific Research and Development. (2024). Drones in Maritime Industry. IJTSRD, 8(4), 520–528. https://www.ijtsrd.com/papers/ijtsrd97093.pdf

Krystosik-Gromadzińska, A. (2021). The use of drones in maritime sector–areas and benefits. *Polish Maritime Research*, *28*, 85–91.

Panico, A., Fragonara, L. Z., & Al-Rubaye, S. (2020). Adaptive detection tracking system for autonomous UAV maritime patrolling. In 2020 IEEE 7th International Workshop on Metrology for AeroSpace (MetroAeroSpace), 539–544. https://dspace.lib.cranfield.ac.uk/items/f5feff39-3e8c-44bd-bfdd-c6d9478911b9

Reports Insights Consulting Pvt Ltd. (2025). Taiwan maritime drone market analysis. https://www.linkedin.com/pulse/taiwan-maritime-drone-market-analysis-key-developments-o6kgf

Sneha, S., & Surekha, P. (2025). Maritime human drone teaming for search and rescue operations. IEEE Xplore. https://ieeexplore.ieee.org/document/10531426/

Verified Market Reports. (2025). Maritime Drone Market Size and Forecast. Retrieved from https://www.verifiedmarketreports.com/product/maritime-drone-market/

KEY TERMS AND DEFINITIONS

AI (Artificial Intelligence): Any computer which does things that appear to be intelligent Nothing More Than a Computer? A.I. facilitates Drones to function and make decisions, independent of human intervention with air navigation, object detection and predictive analytics.

Autonomous: The degree that once started, process need not be monitored. And it is essential to the aspiration of having drones that can act with autonomy, performing surveillance and environmental monitoring or executing security work without human direction.

Coastal Management: The management and regulation of the use of coastal areas to balance environmental, economic, and human needs. In this work UAV's aid to coastal management by mapping coastlines, tracking erosion and other environmental change.

Data Collection: The actual practice of collecting and examining information to find answers about a question or address an issue either applied through research or during routine operations. In this project, drones are mounted with sensors and cameras to monitor up-to-date information regarding maritime conditions, ship activities, and environmental data.

Drone: A remotely controlled or autonomous aircraft used for a variety of purposes, including commercial, military, and environmental applications. In this study, drones are utilized explicitly in the maritime sector for tasks such as port security, offshore wind farm inspection, and environmental monitoring.

Environmental surveillance: systematic monitoring and evaluation of environmental conditions. Here, the drones are used for: Monitoring ocean pollution, tracking oil spills and monitoring marine biodiversity.

Maritime Communication: Ships, ports and notices to mariners communicate naval topics. This is possible through data relayed from drones, monitored radio transmissions and by improved situational awareness to name a few.

Maritime Domain: This analysis is set in the maritime sector and applied on drones used for monitoring, security, logistics, etc. This category includes the shipping and transit related operations. We are concentrating on the domain of maritime in this project, particularly in drone integration.

Mission: A purpose or set of objectives that a particular job, project or organization is intended to accomplish. The work utilizes drones for tasks such as cargo inspection, search and rescue and surveillance.

Offshore Wind Farms: A cluster of wind turbines that produce electricity in marine environments. A drone inspects and monitors offshore wind farms to ensure they are safe and efficient.

Port Security: Efforts to secure maritime infrastructure against smuggling and terrorist activities. In this study, drones aid in port security with real-time surveillance.

Remote Sensing: The process of collecting information about something from a distance, such as with satellites or aerial technology. Scientists deployed drones to collect high-resolution images of sea surface phenomena and data on seawater properties for marine surveillance and research.

Unmanned Aerial Vehicle (UAV): Any flight capable, unmanned aircraft that does not carry a human operator and is controlled either autonomously or remotely Similarly, Piloted Aircraft : An aircraft designed to have a human aboard as the operating agent. An UAV (Unmanned Aerial Vehicle) also called drone, will be used in this work for surveillance, data collection and maritime traffic monitoring.

Unmanned Aircraft System (UAS): The aircraft (drone, or plane), the control station and the communication links as a package of what is needed to run the drone. The technology behind UAS is essential to the efficient deployment of drones in the maritime domain.

Chapter 9
Waves in the Cloud: Rethinking Maritime Infrastructure in a Connected World

Bernandino Malang
https://orcid.org/0000-0001-9926-9461
Bulacan State University, Philippines

ABSTRACT

The maritime industry, long considered the backbone of global trade, is undergoing a transformation as digital technologies reshape how ports, vessels, and logistics systems operate. Traditionally reliant on physical infrastructure and manual coordination, maritime operations now integrate cloud computing, edge intelligence, AI, and IoT to enhance efficiency, safety, and sustainability. This chapter explores how these technologies reconfigure maritime infrastructure into data-driven ecosystems enabling real-time coordination and predictive decision-making. It highlights opportunities such as fleet optimization, predictive maintenance, and smarter port operations, while noting challenges like data sovereignty, cybersecurity, and the exclusion of smaller ports. It stresses that successful transformation must extend beyond technology, addressing governance, inclusivity, and human factors. Ultimately, it argues that a resilient and fair "maritime cloud" can bridge oceans, foster collaboration, and sustain the next era of global trade.

INTRODUCTION

Shipping and trading have always been a cornerstone for global interaction and impacted on the world economy by facilitating trade contact with other continents and carrying goods across the world. From the first sound of a shipyard clang to the last low sound of the cranes that patiently wait in a very crowded port, this field of

DOI: 10.4018/979-8-3373-3546-9.ch009

work has always been a symbol of triumph. As of now, the shipping industry world is going through a transformation that is slow but sure; a change that was not easily accepted. When a task that was handed down is now done through computers or fully manual operated machines. Automation has greatly improved and changed traditional shipping activities into sophisticated, technology driven, integrated activities through all the buzz words; edge analytics, cloud, AI, IoT, and blockchain. Shipping and port system operations today leverage muscle memory, sensors, algorithms, and satellite systems along with real time logistics and synchronization on a global scale. This shift is the distinguishing factor between conventional systems as it fosters a combination of automation and human insight. It is the dawn of intelligent maritime systems.

However, while technology and digitalization are becoming more essential in this modern age, companies still must figure out a way to protect their employees through integration and creating safe workplaces. Many companies that use Software Defined Networks and Network function Virtualization don't realize that the more the company grows, the harder keeping employees safe becomes. Cross country taxes, inconsistent laws, and international laws regarding cybersecurity and data acquisition only complicated the issues. Many companies have shown to use a more hybrid approach to cloud computing along with AI controls and blockchain to maintain company data, while increasing compliance and data resilience through cloud computing.

There is a more important aspect of digital transformation which rests on people and not the hardware or software of a system. Any system still has its operating crew, and its success rests on the people. Maritime professionals and other people need technology designed with usefulness, intuitive interfaces and ongoing training at the cornerstone to help them dilute apprehension or over-alter control. When the focus is on the decision making, safety and trust honors which every rhythm and life at sea stress Foster, the boiling of the sea is no longer a phenomenal feat. Almost every time absence is felt in innovation in any field of its concern, the burden is the concern. The optimal merger of technology and people will determine the smart future of sea work in the digital world. This chapter illustrates constructions of blended, foundational and, in some instances, pioneering development developed every technology, law, and social which step in sophistication encourages the maritime industry toward the construction of a sturdy. Safe, and genuinely smart ecosystem to help steer through the world in its complex ever more tangled web of global trade.

I. THE URGENCY OF DIGITAL TRANSFORMATION IN MARITIME OPERATIONS: MOVING FROM AUTOMATION TO INTEGRATED COORDINATION

For centuries, technology greatly improved maritime activities using things like self-steering ships and almost fully automated unloading cranes. However, the industry is evolving toward more holistic objectives beyond mere isolated automation: full digital integration. Focusing full sorely on machines performing disjointed, standalone tasks is counterproductive. The focus should rather be on formulating an integrated system to allow ships, ports, logistics companies, and governmental departments to digital on a common framework, fully optimizing, interoperating, and exchanging data real-time. Operational data is no longer the focus. The focus should shift to data driven collaborative intelligence using the data as the primary language of international ship trade.

One single instance, such as a vessel changing course, leads to everything happening in a coordinated fashion. Cranes move, vessels take anchor, and simultaneously, customs and port authorities are electronically given base instructions with modified real time data. These chains of activities are executed by the automation of Artificial Intelligence, IoT, and real time data driven analytics. When used collectively, these systems create a network which optimally utilizes resources, increases safety, and decreases delays. All the components of these systems allow maritime institutions to enhance their predictive and prescriptive analytics, speed, uncertainty, boost flexibility, and agility. These actions allow the institution to take advantage of the opportunities presented.

This new set of opportunities comes with significant obstacles as well. Many ports in the world still use the same technology to track and transport goods that they were using thirty years ago. For some reason, the cost to upgrade this technology, so that it matches the global standards to the same degree, is close to impossible to calculate. Despite the resurfacing, some of the technology is far more important than the investment made. There is a serious gap in the availability of information. The market is realizing the importance of information and data to the progress and anticipation of these resources. Such a strategy transitions the emphasis of the industry from control of machines to the coordination of systems, which allows trade across the world to be done unhindered.

From Ships to Data Centers and Ports to Digital Platforms: Cloud Technology as the New Sea Lane of Global Trade

Modern vessels have evolved from simply carrying cargo to functioning as advanced data centers – Acciaro & Sys (2020). These ships gather huge volumes of data to make real time decisions thanks to the interconnected sensors that monitor engine activity, fuel levels, the state of the cargo, and the weather. In the event of an unforeseen problem, such as an approaching storm, these ships can use advanced analytics to re-route to safer and more efficient paths without human oversight. Thus, in this new paradigm, the ship is no longer an inert vehicle, but rather an active node in an integrated maritime network that meshes vessels, ports, and global supply chains (UNCTAD, 2023).

As noted in the scholarly work of Heilig et al. (2017), the old-styled ports have also changed as they are no longer dormant entry points but rather dynamic and analytical platforms. They use real-time and IoT systems to automate and predict when a ship will arrive, then subsequently allocate the ship to workers and equipment, and later track the shipping container from the dock to the storage. The ports are now able to function as digital head command centers to the land and sea logistics simultaneously. The manual coordination functions of the ports have shifted to predictive management as it enables the ports to trade with more accuracy and flexibility.

The use of cloud computing has been reported as the digital maritime network cornerstone (IMO, 2022; Stopford, 2009). Ports and vessels can diagnose and fix problems within schedules, schedules and even rerouting cargo, thanks to cloud-based systems. The cloud's elasticity allows smaller players to purchase advanced analytics and automation tools, without the heavy infrastructure expenditure. As a result, smart ships and intelligent ports, together with cloud-based technologies provide a more adaptable, transparent, and resilient marine environment. In the 21st Century, the strategic maritime sector drives the global economy with information systems technology to improve sector precision, interoperability and cross collaboration (UNCTAD, 2023).

The Need for Robust Digital Infrastructure in Maritime

For many years now, global trade has depended on the sea, with nearly 90% of all goods exchanged internationally shipped across the ocean (Merk, 2018). While still very pointedly important, these facts mask a grim reality; many of the components of the industry are still stuck, outdated, inefficient, and poorly integrated with regard to trading systems. Outdated systems and techniques which have facilitated traditional shipment now struggle with the rapid digital data demands with a world where even the slightest delay can be all it takes to alter global supply chains. As it stands now, the ocean's part in international trade is becoming more vital than ever, which means the sector must transition from a collection of improvisers to a digital, seamless system able to manage speed and complexity.

Ports, vessels, logistics providers, and regulators have never been more critical in the prompt, reliable communication in the digital age. Schoyen and Steger-Jensen

(2017) beliefs, collaboration and the sharing of information are vital and should no longer be considered elective. A mid-route ship receiving real-time port information and slowing down in response, or a terminal that digitally 'filters' cargo movement via predictive algorithms. These examples are actionable, not fictional. They are outcomes of digital integration which are still emerging within the industry.

Real-time dashboards and analytics allow stakeholders to concurrently monitor energy consumption, cargo traffic, and various associated risks. These tools enable maritime stakeholders to respond to disruptions, caused by, for instance, pandemics, geopolitical tension, conflicts, and extreme weather, during which the goal is to minimize downtime. In this situation, data forecasts and manages the entire activity by coordinating control of every supply chain activity. It describes the activity flow within the control center.

Implementing digital strategies will call for enormous investments aimed at the international pool of programmers and engineers. All stakeholders involved, including governments, port authorities, and private operators, need to address the cross-jurisdictional data standard framework to enable collaboration on cross-border interoperability (European Commission, 2020). Even though modernization is expensive, on the other hand, increased efficiency, decreased emissions, and improved supply chain resilience provides long term support to the investment. The current situation of the industry suggests two possibilities; to either stay tied to older and inefficient systems or to adopt change and transform the digital sector which in turn, strengthens the position of the maritime trade in the global economy.

The Role of Government and Regulation in Maritime Digital Transformation

The most critical aspect regarding the completion of the digitization of the maritime sector rests with the government and other regulatory organizations: achieving the digitization in the safest, economical, and ecologically sustainable way possible. They do not only stop at law making; they also advocate new technologies, and provide the appropriate legal foundations necessary to finance and nurture new innovations. In this regard, national administrations can foster competition by shifting to policies encouraging digital competition and lowering new technology investment barriers. Simultaneously, regulatory authorities need to develop comprehensive regulatory approaches to facilitate system interoperability and integration of maritime technologies cross-border and cross-institutions (Cullinane & Haralambides, 2021).

While many countries continue to devote resources toward the achievement of strategies to advance the development of smart port systems and eco-friendly technology, the use and registration of patents, the achievement of all these objectives rests on the functionality of cohesive national and global systems to address the lingering issues of data governance, cybersecurity and system interoperability (Ng, Liu, & fellows, 2018). 'Automating' the maritime industry is hard because data is divided across borders and platforms. Governments need to create and encourage incentives that enhance cooperation between the public and private sectors to improve competition. There is also the need to prevent data breaches and cyberattacks that would jeopardize commercial and international trade, as well as maritime security (Woxenius, 2021). Strong cybersecurity in the maritime operations ecosystem is needed to protect operational data and user relationships.

Additionally, the need to automate the maritime industry is even more difficult due to the global and inter-platform data silos. There is also the need to create well-designed incentives that foster cooperation between the public and private sectors to boost competition. The law still requires that proper cyber security is in place to counter breaches and cyber-attacks that would jeopardize business or international trade and maritime security (Woxenius, 2021). For this reason, the cyber security in the functions of the digital maritime ecosystem is pivotal in data trustworthiness preservation and user's trust retention.

Aside from that, digitalization of the maritime domain is also something that needs to be pursued with proper cyber security in place. It is indeed acknowledged that the application of digital technology can enhance operational efficiency tremendously. However, this should also be paired with energy efficiency and measures of effective green technology.

Along with many governments, the International Maritime Organization (IMO) has taken the lead in promoting the implementation of green digital technologies, such as port electrification, fuel-saving AI-based navigation systems, and emissions-reducing predictive analytic tools (IMO, 2021). This allows the organization to place sustainability at the very core of digital governance policy and ensure that the modernization of maritime operations keeps alignment with global environmental and climate change initiatives.

Cybersecurity in the Maritime Industry

There has been increasing vulnerability to cyber threats, which stems from increasing reliance on digital systems. Networked systems are used in ports and ships for the communication, navigation, and management of cargo, which, while more efficient, are prone to cyber-attacks. In the past, port operations and international shipping have been disrupted because of phishing, ransomware, networked systems attacks, and other such cyber assaults. Such scenarios, however, are highly unlikely with the effect they have on the seamless operation of seafarers, port- based workers as well as on global trade and commerce (Tam & Jones, 2019). It is for such reasons that, for many scholars, the maritime domain is not one which is restricted to the protection of information, but the defense of the entire global supply chain with its various sustaining elements.

To build resilience, it is not only a question of installing sophisticated software but building a culture of awareness as well. The weakest link in any organization is people, even as intrusion detection, firewalls, and encryption form the first line of technological defense. Security breaches are often the result of a lack of hygiene, such as careless handling of classified documents, improper password management, and the like. There is, therefore, need for awareness and training campaigns so that cybersecurity is no longer the sole responsibility of experts. Also, fleets and ports should make it a standard procedure to perform real-time monitoring, system audits, and vulnerability assessments in order to preemptively detect and deter attacks.

Defensive lines do not only rely on relationships with technology. The coordination of shipping lines and port authorities with the government is as vital as any other element. European Union initiatives such as the Maritime Information exchanging and the Analysis Centre (MI-SAC) demonstrate that derived stakeholders are able to stop siloed attacks through information sharing. In this way, cybersecurity is not a question of competition on the market, but a collective shield for the whole industry.

Training is valuable but real security is not only about preventing security incidents. Some cyber occurrences are inevitable. To support redundancy, business continuity must be sustained through contingency frameworks and secure cloud backups sustains critical supply chains even under system compromise.

The International Maritime Organization (IMO) got tougher by adding cyber security to already existing frameworks like the ISM Code. This suggests that cyber security is not an issue that rests with the IT department only; all players in the maritime arena have some responsibilities (IMO, 2021). The maritime sector will, not only embrace, but also ensure that global trade is unobstructed, secure, and reliable.

II. EDGE OF THE OCEAN: CLOUD AND EDGE COMPUTING IN ACTION

Shipping services have made significant advancements with the use of cloud computing. On ships, cloud computing is not only used as a cloud but also as the main storage system combining ships, ports and fleets in real time. Every voyage creates massive sets of information from various sensors such as fuel, engine, weather and navigation sensors. Instead of the desertification of such important data in physical storage, this data also goes to the cloud so that engineering teams and fleet managers are able to access the data from any location. Quick and easy decision making is possible with the data exchange, leading to reduced fuel usage and improved safety in every voyage captured by the UNCTAD in 2023.

An entire fleet's performance can be viewed from one platform thanks to cloud-based dashboards. The ships' performance, along with maintenance schedules and route efficiency, is analyzed in real time, thereby enabling early identification of potential issues (Acciaro & Sys, 2020). For instance, when an abnormal engine vibration is detected, analytics can notify the onboard staff and land-based technicians to avert the problem before it occurs. They also help with predicting weather to determine the best emission and delay mitigating routing as well as altering the vessels heading to real-time weather and sea conditions. Focus in shipping has changed to real-time problem solving instead of event reaction and thinking in hindsight (IMO, 2022).

Cloud computing also helps individual vessels but entire fleets, ports, and entire logistics systems to work together much more easily than before. Better real-time data and processing of the clearance customs, schedules, and cargo reduces inefficiencies in the operations, thus shortening the turnaround time (Heilig et al., 2017). Only ports also take advantage from the automated resource and berth optimization when the shafting vessels traffic altering vessels cross. In this case, the cloud acts as an invisible collaborator enhancing the integration of global shipping through risk and supply chain resilience.

Edge Intelligence: Enabling Real-Time Decisions Even with Limited Connectivity

Even though the vessels rely on the cloud for communication, edge technology makes it possible for the vessels to function autonomously with no connection. As Shi et al (2016) mention, edge computing solves the latency issue with distant computing facilities by processing the computing tasks on the vessel. This is especially true at sea since, for economic and logistical reasons, satellite communication is sparse. A storm is clouds and edge empowered capabilities of priceless, as the system is able to reposition, shift, power, or alter change in velocity—decisions far too critical to be reliant on cloud response—instantly. The ability to change operation without feedback from the cloud is what vessels gain from edge computing. This grants autonomy and increased protection even in the most remote areas (Liu et al, 2020).

Even if some information is not included, edge intelligence fosters improved data handling. For example, ships do not need to send all sensor readings to the cloud. As identified by Liu et al, (2020), ships are able to filter, compress the most important pieces, and only send that. This improves costs, bandwidth, and latency, while onboard real-time processing becomes more efficient. Immediate access to ship diagnostics, predictive maintenance notification, and summarized performance reports is port, while headquarter remote access on minimized reports aids in planning monetarily at the fleet level. Overall, ships are more capable to traverse complex environments and make sound choices.

Satyanarayanan (2017) predicts that the edge cloud systems will integrate along with the fusion of maritime operations clouds. Edge computing, with its autonomy in the ocean, provides speed. On the other hand, cloud systems facilitate fleet-wide coordination and sophisticated analytics. The combination of these systems provides better infrastructure which translates into enhanced safety, reduced fuel consumption, and improved intelligent response of vessels to varying conditions. Increased and more sophisticated autonomous navigation systems as well as more robust vessels will benefit from this.

Why Cloud Alone Isn't Enough and Where Edge Fits In

The advent of maritime cloud computing has changed the operational face of the sector. However, this technological development has brought along some shortcomings. The possibility of relying on cloud-based technology for autonomous decision making, in remote regions with poor satellite coverage, is unsettling (Mansouri et al., 2017). During such crucial instances, the decision is cannot be kept waiting in the machine to be made, as stalled clock cycles waiting for processing power for cloud access can lead to grave consequences in case of repairs and rudimentary environment control. Edge computing solves this issue with cloud service by allowing local to cloud vessel synchronization after processing information and acting unilaterally (Al-Fuqaha et al., 2015).

The concepts of self-governed propulsion systems can be considered to clarify this proposition. The edge enabled system will be able to optimize engine performance and automatic fuel correction even when isolated from shore systems. After reconceptualization of the system, new cloud data will be added to system predictive models, further expanding data from the vessel on fleet models (Islam et al., 2015). The trade-off between decision making and instantaneous action is kept to a minimum, with coordination handled in the cloud and the edge.

This digital partnership is critical for growth of the maritime industry. The edge provides high-speed reliability complemented with global cloud coverage for seamless integration and scaling (Barari et al., 2019). Working in partnership, they develop a flexible digital infrastructure that, while being resilient to the unpredictability of the sea, enables smarter and safer trade.

The Power of Cloud and Edge in Autonomous Maritime Operations

At this point in time, ships that operate on their own are increasingly becoming part of reality because of cloud computing on the edge and the booming technology development of today. Self-sailing ships are operated by radar, sonar, GPS and IoT sensors. The edge cloud boosts system performance computing by route safety and optimization, and edge computing manages real time situational awareness (Shi et al, 2016).

These advances make it possible to use the most cost-effective routes and conserve fuel. Costs operationally are estimated for specific routes using weather, ocean cycles, and the vessel's fuel load, any of which may determine elsewise prohibitively costly operational speeds, for optimum routing and speed on ocean crosses (Poulsen & Sampson, 2019). Like compliance, any savings on fuel are always welcomed. Electronic models of predictive routing, incorporating real-time satellite and port data, predict avoidance of restriction zones which improve the global supply chain and reliable, on-time shipment (Liu et al. 2020).

The devices for edge computing on vessels allow instant practical applications of sensor data. Meanwhile, the cloud uses the historical data collected from many fleets to devise new predictive models, analyzing and deriving patterns from the data. It also enables a hybrid operational framework, where strategic responses to a given problem are both anticipatory and responsive. Predictive analytics, though a technological wonder, is a manifestation of the sector's commitment to safer and sustainable practices. This predictive capability fulfills the increased emphasis on compliance in international trade, which the maritime economy, under global pressure, is in dire need of (Mallidis et al., 2020).

Enhancing Operational Efficiency with Real-Time Data

Before, data operated primarily on a static basis. Now, ships have a multitude of sensors which details the operational parameters of ships mainly on the status of the engine, cargo fuel utilization, fuel consumption, balance, and everything in between. All of which and more, thanks to the cloud, can be computed on the spot. Edge systems can make decision operational in real time, like avoiding storms or fuel consumption. It enhances the ships safety systems and in turn reduces the fuel consumption of the overall operation. Resulting in a double benefit to the economy and the environment.

Shortly, the cloud converts deep local data into top tier advanced strategies. Fleet managers can analyze, organize and schedule all ships in a fleet. Predictive analytics detect fuel or port anomalies and schedule better. The cloud integrates all ships to a worldwide digital ecosystem, providing a basis for intelligent aggregation.

With the improved system, ports also gain something. Geolocated real time data of a vessel's location offers timelier placement of berths and more intelligent placement of inspection cranes. The combination of edge systems and cloud analytics on the entire maritime business provide quick real time decision making and advanced system overall (Rodrigue, 2020).

The Role of Predictive Analytics in Maritime Operations

Predictive analytics has changed shipping from reactive maintenance to proactive planning. Instead of waiting for equipment to fail, sensor data can detect temperature shifts, vibration, or pressure changes that signal early trouble (Acciaro et al., 2018). This approach extends the life of machinery and reduces downtime.

These technologies help logistics planners optimize routes and fuel use. Since fuel constitutes a major portion of operational expenditures, weather conditions, ocean currents, and a ship's cargo are dynamically analyzed to propose a cost-efficient and time-effective route and speed. This observation is certainly understood along with emission and fuel utilization regulations, along with any reduction in fuel consumption which is always welcomed. In the same fashion, predictive routing models utilize satellites in real time and port data to avoid traffic and high-danger regions, assisting the reliability of global supply chains and smoothing and speeding up ocean travels (Liu et al, 2021).

The cloud and edge computing sides merge to create new advances In Technology. Us edge devices on boats do real-time processing on the dynamic sensor data. Simultaneously the cloud works on refining the predictive analytics model and its patterns with historical long-term data collected across fleets (Lee & Lee, 2020). This supports the maintenance of a dual operational system where emerging challenges are reacted to and initiatives are proactively balanced. Therefore, predictive analytics are not only an advancement in technology, but a a reminder of the shift in the maritime industry towards safer, smarter, and more sustainable operations. These strengths in predictive capabilities are in response to the growing international trade compliance and pressures of the maritime economy globally.

III. PORTS, PLATFORMS, AND DIGITAL ECOSYSTEMS

Ports have always had importance in trade, but their roles are changing in the digital age. Smart ports go beyond simple cargo ports and are now part of a connected ecosystem. Smart ports that use the IoT become driven by globalization technology and sustainability. Instead of using big data and machine learning to make decisions, IoT solutions make sensors that can identify and helicopter ports that monitor crane movement as well as port fog. By providing this information, USB ports can identify problems and act before costs are incurred. As such, IoT turns ads ports into "living systems" that can withstand change (Rodrigue, 2020).

Intelligent systems quickly become more sophisticated due to newly emerging technologies. Intelligent systems can predict operational rather than waiting for actions to be taken how to optimize productivity and shipping with tools and transportation. Machines can optimize and coordinate crane dispatch in relation to maintaining order and controlling. AI can predict ship transits with great accuracy, and downtime can be effectively minimized with adequately timed preemptive maintenance. AI and machine learning also enable the prediction and the prevention of lags in operation thus optimizing operational expenditures through minimized fuel usage. Effective cloud computing provides secure and controlled communication among the key stakeholders of shipping lines, customs, forwarding, and even the port. Slack systems replace the fragmented communication and enable improved and faster engagement among the key partners.

Port improvements also have made use of technology for more ecologically sound operations. Systems for smart scheduling lessen emission output and system idling whereas automated environmental systems monitor air and water quality to support greener operations (UNCTAD, 2023). The use of technology and sustainability in harmony shows the responsible port as an evolving image of economic growth. A port that possesses the attribute of smartness is, therefore, more than an agglomeration of digital tools, but an intelligent ecosystem of dispersed, interconnected data, advanced systems, and system-wide cooperation. Through the application of IoT, AI, and cloud computing, these ports have also started to reshape global trade ecosystems into more advanced and ecologically sustainable systems for the future.

API Economies and the Platformization of Maritime Services

The usage of APIs is integrating the maritime industry digitally and improving the modalities used by ports, shipping lines, logistics companies, and customs officials. In the past, there were siloed, self-contained systems which endemic errors, delays and increased costs (Jovic et al, 2019). There is a solution offered by Application programming interfaces, or APIs. It enables disparate digital systems to interact. One example is a shipping line's direct interlinking of its schedules to the port's berth allocation system. Also, customs and logistics systems can exchange clearance information in real-time. Rapid integration improves the efficiency of cargo handling while enhancing the ease of distribution in cross border supply chain.

APIs also make it easier to promote the wider "platformization" of maritime services. Digital platforms that dot shields ports and shipping companies enable simplified booking, tracking, customs, and consolidated paperwork within an integrated seamless interface L et al. 2020. These systems are similar to e commerce or banking marketplaces which empower participants to exchange and trade information and services seamlessly. For instance, a logistics provider can be given a single system with tools via APIs for booking cargo space, tracking a shipment, and performing compliance checks. This modularity allows new services to be added in parallel to running operations and enables contemporary innovations to thrive unhindered and also allows for the pedestrian industry to pivot to services (Hofmann & Osterwalder, 2017).

In addition to efficiency, resilience, and sustainability which were primary objectives of the covid 19 period, became significant attributes of operations systems. Remote port operations which were enabled using the nad technology integrated work systems coordinated by APIs (Notteboom & Haralambides, 2020). Today, APIs interconnect systems for maritime, energy, and fuel optimization, and promote global climate change goals by emission reduction (Poulsen et al., 2021). Maritime organizations that integrate APIs are bound to derives significant operational flexibility which ultimately translates to the entire industry being transformed to a cohesive digital network wherein the services and data transit along with the goods seamlessly.

The Challenge of Digital Inclusion: How Small Ports Catch Up

Digitalization does make international trade much smoother and easier than it was before, however, it does widen the disparity between larger and smaller ports. Regional ports lack the sophisticated tools and frameworks, as well as the money and polished, adept workforce in comparison to the major ports such as Rotterdam, Singapore, and Shanghai, which spend millions on advanced AI scheduling, IoT automation, and automation tracking (Ng & Cetin, 2020). This creates an imbalance in the economy by risking a maritime economy, where one economy is modern and advanced while the other is manual. In many cases, it is not a lack of ambition, but many small ports have to experience a major financial burden to fully digitalize (Merk & Notteboom, 2015).

With affordable and scalable solutions, Cloud computing assists in bridging the digital divide by decreasing the need for expensive local infrastructures. Thanks to the cloud, smaller and mid-sized ports can afford to monitor and control the movement of ships as well as manage the logistics and customs files in real time (Heilig et al., 2017). The "cloud" system automates the ports to a sufficient level enabling them to modernize core functions to adapt to advanced frameworks over time. Further advantages through shared training and expertise, technology adoption, and training facilitates the integration of large and midsized ports. Regional "port communities" helps smaller ports which lack the sophisticated tools on their own and strengthens the competitiveness by pooling resources.

The collaboration of different policies is just as important. Gov't initiatives and the global community as exemplified by the IMO's e-navigation project, assist and facilitate small ports' modernization and active participation in digital trade. Schoyen & Bråthen, 2021. The e-navigation project strives to assist small ports in the digital marketplace while ensuring the environment remains protected at the same time. Such harmonization of initiatives fosters "Inclusive Green Growth." As Rodrigue mentioned in 2020, the e-navigation project fosters the belief that the world economy is "growing more global, smaller, and faster," and that more ports and regions must be connected and accessible in the fundamental digital economy to remain. The economy circulates faster and becomes stronger with more ports connected.

The Role of Data Sharing in Building Collaborative Port Ecosystems

Currently, the smart ports of the world are sustaining collaboration on the basis of data sharing. Before, the data residing within different shipping lines, customs, and terminal operators was very siloed and fragmented, and, as a result, the logistics and the entire process was inefficient and lacked transparency. Smart Ports act like a smart and connected system, enabling data to flow across different stakeholders. If shipping lines provide ports with real-time shipping arrival estimates, ports can manage crane and berthing capacity and customs can prepare the documents. This used to be very different workflows which are now styled processes saving time, reducing congestion, and cutting costs.

Real-time data sharing fosters open data platforms, which, in turn, fosters accountability, transparency, and trust. When ship operators, other logistics providers, and the authorities are on the same page with real-time data, the possibility of disputes over cargo and delays is eliminated. With improved data sharing, even accountability and trust improve. During the COVID-19 pandemic, ports with consolidated and integrated systems data adapted, rerouting shipments and reallocated resources with no disruptions. All of this is focused on resilience, especially concerning climate change, risk increases from cyberattack, geopolitical tensions etc. This type of resilience being a competitive advantage in the market is what Liu et. at refer to (2020).

How Digital Twins Are Revolutionizing Port Operations

One of the newest technologies that help with port management is called a digital twin. A digital twin is a data-enabled working replica of physical objects, as opposed to a digital model static in time, like cranes, terminals, berths, and warehouses, with constant updates through data analytics and IoT monitors. They assist operators in visual analysis, forecasting, and real-time optimization (along with some other things). A digital twin, for example, might be able to compute changes in equipment condition to yank cargo-flow bottlenecks and mitigate them in advance to prevent any disruption. This transition from disrupted to disrupted-focused management is a big leap that ports around the world are embracing (along with wisdom, of course).

Digital twins are equally and more effective at maintaining efficiency. Systems can make projections based on recorded metrics in reasonable time cycles (such as vibrations, uses of energy, and other things), and this prevents expensive downtimes. It is even possible for them to congestion and suggest other routes that are more efficient for constant container movements. Even more, managers can preemptively design simulations of randomly assembled real ports. This way, the ports can to some extent partake in safe experimentation. This sustains tethered evidence management, and lowers the risks in decision making, which is better control for the world's ports (evidence management, rather than proof).

Besides efficiency, digital twins also promote sustainability and collaboration. They allow the testing of green initiatives like electrification and emission-reduction route planning (Kusic et al., 2023). When combined with the digital twins of shipping and logistics companies, they create an ecosystem with cross supply chain visibility (Callcut et al., 2021). This integration supports better collaboration and the management of energy, compliance, and logistics. Ultimately, hops with digital twins will be more agile, data-centric, and prepared for the evolving trade landscape.

The Environmental Impact of Smart Ports

The use of digital technology is changing sustainability focus areas for maritime operations. What were once major emission sources from the ports are now being mitigated using smart technologies to lower environmental impact (Acciaro et al. 2018). AI-led scheduling along with IoT driven control systems manage cranes and vehicles to minimize delays to improve fuel efficiency. Better allocated yard operations reduce the number of container moves dropped, resulting in lower energy use and greater throughput (Lam and Notteboom 2014).

The improvement in energy use and consequent reduction in air and water contamination is a major benefit emerging from these changes. Ships permitted to dock and plug into the shore power system instead of running their diesel engine, steaming into dock. Evidence from ports like Rotterdam and Los Angeles demonstrates the systems to cut nitrogen-oxide and particulate emissions on the water and in the surrounding atmosphere (Poulsen et al. 2018). Environmental IoT sensors rigged in the atmosphere, water, and noise in real time, and digital twin systems visualize impacts of changes on sustainability before implementation (Callcut et al. 2021).

Collaboration is important to keep this progress going. Port authorities, shipping firms, and the regulators can use shared data systems to track compliance with promote low-carbon, and global sustainability goals, including the UN SDGs and the IMO targets for decarbonization. Innovation with responsibility is just like technology with the smart ports. They provide new, resilient economic gateways which are environmentally sustainable. (Yau et al., 2020).

IV. RETHINKING CONNECTIVITY: SDN AND NFV FOR THE MARITIME WORLD

For many years, communication at sea has regarded the strength of a connection, the stronger the link the better the connection.' Even though satellites enabled global coverage and constant connection for ships, those days are approaching the end of the line as the world digitizes. Ships are now able to stream large quantities of data from their IoT sensors, weather analytics, and crew communication, all of which are vying for restrictive bandwidth. Even though congestion and inequitable data prioritization are issues that come with an increase in signal strength, none of those issues would be resolved. The industry is starting to embrace Software-Defined Networking (SDN). This software is programmable and controls the flow of information in complex seafaring systems (Guesmi et al, 2020).

SDNs enhance the efficiency and flexibility of communication services by allowing the operational and control layers to function separately. As a result the operators can dynamically change the bandwidth allocation, change the paths of the data flows, change the priority of the data streams, and provide priority to data streams, and provide priority to time-sensitive data like navigation advisories or emergency alerts depending on the data conditions on the network. For example, in the instance of a storm that weakens the satellite connections, SDN can change the routes of the vital communication automatically to alternative routes, preserving the necessary communications and safety operational. Programmable networks improve the efficiency and resilience of the network. The coordination of the vessels and the port operations for logistics and maintenance reporting, and auto scheduling are smoother. The systems can prevent the disruptions and preserve the reliable and open global trade routes (Wang et al. 2017).

Due to SDN's network architecture, asynchronous updates to SDN-enabled device systems are possible without physical replacements of machines or expensive new hardware, which promotes maritime communication sustainability. SDN also promotes energy efficiency by optimizing traffic via real-time routing or speed alterations to minimize fuel consumption and emissions (Mao et al., 2016). As more digitized platforms are integrated to maritime networks, reliance on potent connectivity for maritime communications will change to dependence on adaptive, intelligent SDN systems for more streamlined interoperability and self-optimization within new global maritime digital networks.

SDN as the Conductor of Data Flow at Sea

Data is needed just as much as fuel at sea. It is used in navigation, tracking the cargo, and communicating with other vessels and at the ports. But outdated communication technologies which have inflexible and hardware-defined channels, and are used in the sea communicating systems, cannot cope with the huge volumes of smart vessels data and other IoT systems. Operational critical data gets disrupted when multiple data streams try to use the same bandwidth. To fix this, many fleets use Software-Defined Networking (SDN) which helps manage and prioritize data flow.

SDN changes the maritime sector networks. By decoupling the control and data planes, it allows real-time management of traffic and centralized control at the sea subdivisions (Guesmi et al., 2020). For example, SDN can guarantee alterations in navigation and safety command updates get pushed out during a storm, while other data systems like performance reports get pushed down. This ability to prioritize is important during maritime missions where the use of bandwidth changes rapidly. SDN improves communication at sea, guaranteeing systems critical to safety and control systems over the other transient control systems. SDN allows the communications system on a ship to be a self-organizing system. It not only delivers data but also controls the flow of data in the system.

One of the best things about Software-Defined Networking (SDN) is avoiding congestion on the network and scaling the network to fit user's needs. SDN analyzes the data streams and adjusts the data routing to mitigate bottlenecks. This is like how boats are rerouted before they reach over congested ports (Cervelló-Pastor et al., 2017). This helps SDNs keep operations running and avoid sluggish performance. Advanced tools like digital twins and autonomous vessels or AI predictive analytics hinge on the SDN to keep maritime networks scalable and flexible avoiding costly hardware overhauls (Mao et al., 2016).

SDN most importantly changes how maritime communication is configured. It moves and coordinates the disparate parts of the system to a digital ecosystem. Like a conductor who brings an orchestra to organizational and performance unity,

SDN coordinates disparate parts of the information system to harmonize and drive the flow of information vital to global shipping.

V. WHO OWNS THE WAVES?

Real-time data sharing and processing by ships, ports, and logistics operators over cloud computing is game changing. Coordinating global efforts is seamless, but data sovereignty, privacy, and regulatory issues come along with it (González Fuster, 2019). Vessels navigating different countries' legal environments while processing cloud data containing navigational routes, cargo manifests, fuel logs, and cargo details may be challenged. The General Data Protection Regulation (GDPR) compliance may be stringent in the European region, but there are others with loose and undefined borders with even no compliance. For an cross border operating industry, the variations in legal compliance caused by border controls are the biggest governance hurdles.

Privacy concerns vex operational efficiency. Maritime operators rely on raw, unrestricted data flow to monitor their fleets while government agencies want to impose data barrier controls to mitigate sophisticated cyber-attacks and to protect their interests. The 2017 ransomware attacks on Maersk are prime examples of poorly managed data controls in the maritime industry and highlight the risk of cyber-attacks (Jovic, 2020). To have resilient maritime systems, speed and operational flexibility should be allowed in data managed systems while compliance systems are put in place to protect sensitive data.

Companies are using hybrid models for data governance. In these frameworks, sensitive data is stored on local servers, but non-sensitive operational data is allowed to cross borders. This approach complies with national data regulations and keeps operational analytics aligned for fleet systems. Long-term, however, progress relies on global harmonization, developing a common approach maritime data governance to mimic the work the International Maritime Organization (IMO) does to standardize maritime safety and navigation. Without consistent global frameworks, the industry will experience the negative effects of fragmentation, stifling innovation and restricting trade. Most importantly, the maritime industry needs to ensure its data, and cargo, are able to cross borders seamlessly and legally (Poulsen et al., 2018)

Data Governance Models in Maritime Operations

Digitalization in the maritime industry poses new challenges in terms of the management and governance of data. Guns, ports, and logistics companies capture, record and stores large scale data on navigation, cargo, and cargo predictive maintaining data logs. Data management takes quite a lot of effort to compete, security and compliance issues, frameworks and navigation are required. There are three governance models to choose from: centralized, federated and hybrid. Deciding on the "best" model to follow will depend on the additional benefits and drawbacks in each and every model.

The centralized model is the easiest and simplest of all governance models. The data is centralized to one account and control is given to one cloud service. This is a powerful governance model, as commandeers complete control of their dashboards and can operate and analyze data in real-time. Take, for instance, a shipping company that is given complete dashboards to monitor the state of their entire fleet and analyze their fuel consumption. The model does have drawbacks, though. Keep in mind that whole control and all data being uploaded to one server and one cloud model is a risk in itself. In extreme cases, a governance model can entirely eliminate the control of data.

The federated model is, in part, the opposite in many aspects of the centralized model. In a federated model, control and power of data governance is shared with international ports and regional operators, and local governance is implemented. This creates new issues with governance, as deferring the management of control data to local authorities lowers a system's governance model. The model does have benefits: the risk of a system's data being utilized for extreme governance control is mitigated.

The hybrid model combines both strengths. Sensitive data (such as crew identities and security logs) is stored locally to comply with privacy legislation. Less sensitive data is shared locally and globally for coordination and analytical purposes. This approach meets legal requirements while also providing operational flexibility (Taddeo & Floridi, 2018). Hybrid systems offer advanced cybersecurity as well: centralized systems may possess stronger security features technology-wise, but they are far bigger and easier targets. In contrast, federated systems are more secure in totality, but they are way more difficult to breach and standardize. Given that the sector accounts for more than 80% of global trade by volume, hybrid governance is the most sensible option for fostering innovation, compliance, and flexibility within the industry (UNCTAD, 2023; Manyika et al., 2021).

Balancing Privacy and Efficiency in Maritime Cloud Systems

Switching to cloud-based systems has changed how the maritime sector conducts trade, but it has also created new challenges regarding privacy. While the GDPR requires data to be handled and transferred within the boundaries of the law, maritime data operations must also be fast and safe, and the data must cross borders to be efficient. Voigt & Von dem Bussche 2017. For example, an Asia to Europe crossing vessel has to send and perform navigation data to cloud servers in several different countries. By real-time monitoring, the data cloud servers have stored, and navigation data are still being crossing borders, the cloud servers have stored, and navigation data are still being crossing borders, the cloud servers have stored, and navigation data. This creates an absurd situation of legal and logistical challenges.

New systems being deployed in the privacy of maritime trade focuses on speed. While data transferred and encrypted is data-less, blockchain technology is also used to create decentralized and transparent records and to be used to an encrypted and decentralized storage collar. Most cloud storage systems perform pseudonymization to render data-less and stripped of personal details and then perform analysis. This privacy system allows the user to predetermine the detailed hetroot analysis. Other examples are maintenance predictions and also eluding data on trends of covering during a trip.

The future involves a mix of different ways of governing and global alignment. Hybrid systems protect personal information at a national level while enabling the safe transit of anonymized, aggregated data across global networks. At the same time, the IMO and UNCTAD are collaborating to formulate guidelines that reconcile international trade with domestic legal systems. This kind of collaboration, from a stakeholder relationship perspective, bleeds compliance burden and fostering trust. The global cloud ecosystem of the maritime industry will thrive on digital transformation only if it is able to strike a solid balance between global efficiency and privacy.

Regulatory Barriers to Global Data Sharing

The quick increasing use of digital technology in the shipping industry makes it especially difficult to comply with the boundaries of global maritime law when sharing data. Operational data from ships and ports relying on the cloud for things like predictive analytics, fleet management, and real-time logistics crosses more borders with more diverse, conflicting, and technology complicating each jurisdiction. This law and technology gap continues to get worse and more tangled. China's strict data policies coined as "data localization", claim all operational data must be contained within the geographical boundaries of China. While places like Southeast Asia, have considerably more relaxed data policies (Chander & Lê, 2015). This kind of perplexing unpredictability further decreases efficiency for shipping companies as the cost of compliance continues to rise.

The main focus here is in regards to data sovereignty which views data as an asset to the country. These nations spend substantial resources on data cybersecurity and national security which is why there are strict policies in maritime data such as cargo manifests, vessel movements and port schedules and where they are stored and how they are shared and used. These policies are created to protect the national interests; however, they greatly fracture digital ecosystems forcing companies to maintain individual siloed on-premise databases rather than leveraging the shared cloud (Manyika et al. 2021). This fracture exists relative to real-time communication as information provided by vessels to headquarters can be stalled or in some cases data can be missing. Decision making gets impaired, costs increase, and competitiveness is lost by operators as they work under multiple jurisdictions (UNCTAD 2023).

Experts have asked the International Maritime Organization (IMO) and other institutions to create and consolidate data governance frameworks. These frameworks would provide common standards for data encryption, transfer, and compliance, as was the case with e-navigation (IMO, 2021). Until that time, most companies use cloud vendors that are geographically agnostic to regulations. The status of the maritime industry's digital backbone will depend on the degree to which sovereignty, efficiency, and global coherence will remain a strong balance.

Technology Solutions for Data Sovereignty Challenges

New technology is enabling new ways to secure data and facilitate compliance with international regulations for maritime industries. Some of the new tech includes blockchain along with hybrid cloud architecture and AI. The toolkit offers AI-driven data management as well. Sophisticated and tamper-proof transaction records are created with blockchain. By providing real-time audit logs for cargo manifests and customs clearances fuel logs and manifests, fuel logs, and customs clearances, trust and traceability are automated, and freight audit logs are created. The self-contained blockchain network makes compliance within bordered data controls possible, eliminating the need for a data facility back office. Local borders contain sensitive data.

Cloud environments provide secure data storage in hybrid cloud environments, combining the benefits of public and private environments. For weather or performance tracking, public clouds can be used to perform cloud analytics. For sensitive analytics, private local servers can be used. Self-learning AI systems execute compliance with regulations by automatically classifying, encrypting, and monitoring the data transfers that flow between multiple countries. Emerging data compliance technologies like homomorphic encryption allow the rule of data raw ie contents exposed and the final contents encrypted. Compelled compliance technologies are the best to have around.

International coordination due to the IMO's continued support for such innovations is building a more secure and interoperable marine data digital ecosystem. These innovations, with the elements of trust, automation, and flexibility, remain confident that digital transformation will reinforce global trade and not undermine it.

The Role of International Organizations in Global Data Governance

As ships, ports, and logistics companies gather more and more digital data, the need for coordinated cross governance has grown. This is evident in the global standard-setting initiatives on data privacy, cybersecurity, and operational integrity spearheaded by the International Maritime Organization (IMO), the United Nations (UN), and the European Union (EU). Such international cooperation is pivotal in the absence of which inconsistent, divergence, and conflicting national policies on trade will stifle innovation and trade network thrivinganship.

Having previously centered on safety and environmental regulations, the IMO is now integrating digitalization and cybersecurity in its focus. The 2021 guidelines on managing cyber risks at sea represented a significant milestone in the alignment of global practices. The introduction of the EU GDPR has also influenced global maritime operations in the EU's borders by establishing global principles of transparency, accountability, and privacy-by-design. Through UNCTAD, the UN also fosters cross-border data policy discussions, citing fragmented policies in data collection and of its use directly works against efficient supply chain management (UNCTAD, 2023).

Along with legal frameworks, attention to technical standardization is fundamental. Global collaboration on encryption, interoperability, and access control, similar to ISO standards for quality management, can enhance the safety and efficiency of worldwide data exchange (Alcaraz & Zeadally, 2015). Confidence and information misuse is mitigated with public-private partnerships facilitated by confidence providers. Integration of the IMO, EU, and UN will be necessary to guarantee secure, efficient, and cooperative progress on the digital transformation of the maritime industry.

VI. HOW CREW USE MARITIME PLATFORMS AND WHY USABILITY IS IMPORTANT

From navigation boards to communication apps, digital platforms are vital in maritime operations. The simplicity of a technology in use outweighs the complexity of the functions it provides. Even the most sophisticated system will fail at sea if it is too complicated for the crew to understand. Valuable seconds may be lost during a confusion-induced delay while vital hidden information may prompt faulty decisions to be made. Usability is more than a design preference in this case; it is a necessity. Poor design and malfunctioning technology can have dire consequences.

Design firmly crouches at the center of providing confidence in shipping operations. Friction is reduced when systems are reliable and allow for routine, seamless information access. Usable systems require fewer steps to be taken, and more information is easily processed and acted upon when it is needed, especially in the performance of maritime operations. Systems are subjected to extreme maritime conditions; technology that withstands the elements will be a helpful interface at sea. Systems that provide friction less interfaces will provide the confidence needed and clarity in the performance of shipping operations.

Usability is also a reflection of human diversity. The crew of a ship may come from different parts of the world with different levels of digital proficiency. Systems designed with multiple languages and visual signs and help features built into the systems aid in the adoption of these systems. If seafarers feel supported and not overwhelmed with the digitized systems, they will trust and use these systems consistently. Usability boosts safety, productivity, and morale for ship operators. Highly digitized systems will only work if the users trust them to function reliably at sea.

Ensuring Digital Infrastructure Supports Human Decision-Making

Due to technological advancement, the maritime industry has greatly automated its operations, analyzed data, and included AI in its functions. Still, technology should augment the judgement of a human and not replace it. While machines can analyze tons of data, experience, gut feeling, and consciousness can only be offered by a mariner (Costa et al., 2017). An example of this is a route-optimization system, which may suggest the most efficient shortest path, but this would be more fatiguing to an already exhausted crew, and a captain would not take this route. Human oversight is critical for safety more than anything else, however, it is also needed for other decisions. One of the most notable ones is the issue of transparency. There should be a balance between reliance and rejection. To rely on means to trust an automated system to the point where no human judgement is used, and to reject is to not trust it at all. Both scenarios are dangerous for operational safety.

Culture and Education Impact Seafarers' Digital Adoption. Most seafarers come from background where seamanship is learned from practical knowledge. If tools are introduced too quickly, there will be some backlash. There is a specific type of training that focuses on technology as a supporting partner—not a substitute— to fill this gap. Success in diverse maritime context depends on the integration of data computing and human contribution, focusing on data realism and context of the decision.

Integrating Human Factors into Maritime Platform Design

Success in the digital age means more than just mastering technology. For people to be satisfied, technology has to work seamlessly with their human strengths and weaknesses. Human factor engineering ensures that platforms 'work' in the world of ships and boats, where fatigue, fatigue, and concentration stressing motion are constant challenges (Grech, Horberry, and Koester, 2019). Systems built with these in mind become trusted aids to judgment and safety.

Systems makers primary focus has to be the cognitive load, or the amount of mental effort a user has to use to make sense of and use the information available. Disorganized interfaces that offer too much information can 'flood' users and make them slow to act. Comprehensible systems place in the hierarchy of things a premium on touch and goes collision alert systems or engine fails (Stanton et al. 2017). Also important are the physical aspects of the systems - the ability to touch devices, the ability to change the globe and weather and mechanical conditions.

Emotion and situation perception also pertain to design. While Hetherington et al (2006) have noted that crew members working extended periods of time, alone and isolated, are under stress and need calming visual design, guided feedback, and unambiguous 'go' messages for control, adaptability across zones (from the bridge to the engine room) and across devices provides an integrated and cohesive experience. Having seafarers as active participants in usability testing and design cycles ensures that systems remain current and relevant as workflow changes with time. To put it simply, human-centered design shifts how technology is perceived. Instead of thinking of it as just another digital device, it is now viewed as a reliable operational partner that enhances safety and performance.

The Role of Training and Support in Digital Integration

Technological innovation in shipping is effective only when paired with human expertise. Use of advanced cargo handling and control systems would depend on crews' ability to understand the systems and logic advanced technology employs. Digital integration begins with training and continuous assistance. As the tools designed digital tools integration might be sensible, poorly envisioned assistance gaps will generate frustration and misuse– tools will be better designed gaps will still be frustration and misuse as tools will be poorly designed.

Within the framework of the programs, the candidate's technical ability and the ability to make sound decisions simultaneously should be built. Crew members need to understand the system's ability to process data and when and when the system should make the decision to overrule the system Barnett et al., 2006. Specialized preparation, designed to be responsive to change, systems self-writing analysis and updating, and self-dialing and diagnostics, as well as self-maintaining systems predictive systems, as well as self-dialing and diagnostics systems, can be grouped as primary systems as well as crews on boards, field support, and peer assistants to remote support systems– engineers focus on self-maintaining advanced technology.

Having a strong mind is just as crucial as the rest. Most seafarers who always worked with manual systems may initially be skeptical to the notion of automation. Training focuses on how digital instruments improve safety and lessen workload in order to build confidence and trust. Companies must also view training as a prospective investment and routinely revise it with the development of new technologies. If skill development and technical support fostered together with human engagement, the ease of adoption is seamless. The improvement in the process is bound to result in a positive shift of performance and resilience in maritime safety.

VII. CONCLUSION

The use of advanced technology to transform the maritime industry signals a shift toward smarter and more connected data base. Still, as the above discussion described, its success does not depend only on technologies but absence of advanced technologies and the level to which these innovations augment human capabilities and discretion. Artificial Intelligence, the internet of things, cloud computing, blockchain, and software defined networks are changing the way we navigate, manage safety and security, and the way we conduct logistics. These system's capabilities are only realized when the system is designed with the people and not the people as the primary focus.

Systems that incorporate human-centered design are more likely to have people anchored in the operations of the system. When members of the maritime systems have digital platforms to work with that are available, flexible, and responsive to the current maritime environment, not only does better coordination occur but also the promotion of a safety culture and trust. Thus, the systems designed are the ones that augment human judgement and not replace them, that is, human discretion and judgement are bound to the center of the system.

That said, every sophisticated piece of technology, without the combination of automation and human integration, is of little value. It is given that machines can retrieve, process, and analyze data faster and more precisely than any group of people. But machines don't possess the ability to intuit, appreciate ethical principles, and understand the context. This is why system design must be transparent. This allows system operators to understand, question, and change automation when the situation requires. This persistence of automated digital feedback systems is what makes digital transformation humanized. It changes and grows with its operators because of the ever-evolving human factors engineering and the people.

In the same manner, there is a gap of system autonomy which automation and ongoing assistance can fill. Digitalization is never an endpoint. It is an evolving process. Hence, commitment of resources for long term learning, flexible retraining systems, and responsive system maintenance is a must. This way the new generation of maritime professionals has the opportunity to defend and take charge of the new technology without compromising the safety and efficiency of the operations.

To wrap up, technology, human capital, and continuous learning integration will determine the future of maritime operations. When innovation and usability develop in parallel, the outcome of an ecosystem is efficient, sustainable, and people-centered. Thanks to such people-oriented integration, the industry can expect optimistic digital transformation and safe, collaborative, resilient oceans.

REFERENCES

Acciaro, M., Ghiara, H., & Cusano, M. I. (2018). Energy management in seaports: A new role for port authorities. *Energy Policy*, *71*, 4–12. DOI: 10.1016/j.enpol.2014.04.013

Acciaro, M., & Sys, C. (2020). Innovation in the maritime sector: Aligning strategy with outcomes. *Maritime Policy & Management*, *47*(8), 1045–1063. DOI: 10.1080/03088839.2020.1737335

Al-Fuqaha, A., Guizani, M., Mohammadi, M., Aledhari, M., & Ayyash, M. (2015). Internet of Things: A survey on enabling technologies, protocols, and applications. *IEEE Communications Surveys and Tutorials*, *17*(4), 2347–2376. DOI: 10.1109/COMST.2015.2444095

Alcaraz, C., & Zeadally, S. (2015). Critical infrastructure protection: Requirements and challenges for the 21st century. *International Journal of Critical Infrastructure Protection*, *8*, 53–66. DOI: 10.1016/j.ijcip.2014.12.002

Barnett, M. L., Gatfield, D., & Pekcan, C. (2006). A research agenda in maritime crew resource management. *Proceedings of the Human Factors and Ergonomics Society Annual Meeting*, *50*(19), 2081–2085. https://scispace.com/papers/a-research-agenda-in-maritime-crew-resource-management-14o7j2f2uw

Callcut, M., Agliozzo, J. P., Varga, L., & McMillan, L. (2021). Digital Twins in Civil Infrastructure Systems. *Sustainability (Basel)*, *13*(20), 11549. DOI: 10.3390/su132011549

Chander, A., & Lê, U. P. (2015). Data nationalism. *Emory Law Journal*, *64*(3), 677–739. https://ssrn.com/abstract=2577947

Costa, N. A., Holder, E., & MacKinnon, S. N. (2017). Implementing human-centred design in the context of a graphical user interface redesign for ship manoeuvring. *International Journal of Human-Computer Studies*, *100*, 55–65. DOI: 10.1016/j.ijhcs.2016.12.006

Cullinane, K., & Haralambides, H. (2021). Global trends in maritime and port economics: The COVID-19 pandemic and beyond. *Maritime Economics & Logistics*, *23*(3), 369–380. DOI: 10.1057/s41278-021-00196-5

European Commission. (2020). *Europe's digital strategy: Shaping Europe's digital future*. European Commission. (2020). *Europe's digital strategy: Shaping Europe's digital future.* https://digital-strategy.ec.europa.eu/en?utm_source=chatgpt.com

González Fuster, G. (2019). *The emergence of personal data protection as a fundamental right of the EU*. Springer., DOI: 10.1093/idpl/ipu034

Grech, M. R., Horberry, T., & Koester, T. (2019). *Human factors in the maritime domain*. CRC Press., DOI: 10.1201/9780429355417

Heilig, L., Lalla-Ruiz, E., & Voß, S. (2017). Digital transformation in maritime ports: Analysis and a game theoretic framework. *NETNOMICS: Economic Research and Electronic Networking*, *18*(2–3), 227–254. DOI: 10.1007/s11066-017-9122-x

Hetherington, C., Flin, R., & Mearns, K. (2006). Safety in shipping: The human element. *Journal of Safety Research*, *37*(4), 401–411. DOI: 10.1016/j.jsr.2006.04.007 PMID: 17046789

Hofmann, E., & Osterwalder, F. (2017). Third-party logistics providers in the digital age: Towards a new competitive arena? *Logistics (Basel)*, *1*(2), 1–28. DOI: 10.3390/logistics1020009

International Maritime Organization (IMO). (2021). *Guidelines on maritime cyber risk management*. https://www.imo.org

Islam, S. R., Kwak, D., Kabir, M. H., Hossain, M., & Kwak, K. S. (2015). The Internet of Things for health care: A comprehensive survey. *IEEE Access : Practical Innovations, Open Solutions*, *3*, 678–708. DOI: 10.1109/ACCESS.2015.2437951

Kusic, K., Schuman, R., & Ivanjko, E. (2023). A digital twin in transportation: Real-time synergy of traffic data streams and simulation for virtualizing motorway dynamics. *Advanced Engineering Informatics*, *55*, 101858. Advance online publication. DOI: 10.1016/j.aei.2022.101858

Lam, J. S. L., & Notteboom, T. (2014). The greening of ports: A comparison of port management tools used by leading ports in Asia and Europe. *Transport Reviews*, *34*(2), 169–189. DOI: 10.1080/01441647.2014.891162

Liu, Z., Schindler, S., & Liu, W. (2020). Demystifying Chinese overseas investment in infrastructure: Port development, the Belt and Road Initiative and regional development. *Journal of Transport Geography*, *87*, 102812. Advance online publication. DOI: 10.1016/j.jtrangeo.2020.102812

Mansouri, Y., Toosi, A. N., & Buya, R. (2017). Data Storage Management in Cloud Environments: Taxonomy, Survey, and Future Directions. *ACM Computing Surveys*, *50*(6), 1–51. Advance online publication. DOI: 10.1145/3136623

Manyika, J., Lund, S., Bughin, J., Woetzel, J., Stamenov, K., & Dhingra, D. (2021). *Digital globalization: The new era of global flows.* McKinsey Global Institute. Retrieved from: https://www.mckinsey.com/capabilities/mckinsey-digital/our-insights/digital-globalization-the-new-era-of-global-flows

Mao, H., Alizadeh, M., Menache, I., & Kandula, S. (2016). Resource management with deep reinforcement learning. In *Proceedings of the 15th ACM Workshop on Hot Topics in Networks* (pp. 50–56). ACM. https://doi.org/DOI: 10.1145/3005745.3005750

Notteboom, T., & Haralambides, H. (2020). Port management and governance in a post-COVID-19 era: Quo vadis? *Maritime Economics & Logistics, 22*(3), 329–352. DOI: 10.1057/s41278-020-00162-7

Rodrigue, J. P. (2020). *The geography of transport systems* (5th ed.). Routledge., DOI: 10.4324/9780429346323

Satyanarayanan, M. (2017). The emergence of edge computing. *Computer, 50*(1), 30–39. DOI: 10.1109/MC.2017.9

Shi, W., Cao, J., Zhang, Q., Li, Y., & Xu, L. (2016). Edge computing: Vision and challenges. *IEEE Internet of Things Journal, 3*(5), 637–646. DOI: 10.1109/JIOT.2016.2579198

Stanton, N. A., Salmon, P. M., Rafferty, L., Walker, G. H., Baber, C., & Jenkins, D. P. (2017). *Human factors methods: A practical guide for engineering and design* (2nd ed.). CRC Press., DOI: 10.1201/9781315587394

Stopford, M. (2009). *Maritime economics* (3rd ed.). Routledge., DOI: 10.4324/9780203891742

Taddeo, M., & Floridi, L. (2018). How AI can be a force for good. *Science, 361*(6404), 751–752. DOI: 10.1126/science.aat5991 PMID: 30139858

Tam, K., & Jones, K. (2019). Cyber-risk assessment for autonomous ships. *Reliability Engineering & System Safety, 195*, 106698. DOI: 10.1109/CyberSecPODS.2018.8560690

United Nations Conference on Trade and Development (UNCTAD). (2023). *Review of maritime transport 2023.* United Nations. https://unctad.org/webflyer/review-maritime-transport-2023

Voigt, P., & Von dem Bussche, A. (2017). *The EU General Data Protection Regulation (GDPR): A practical guide.* Springer., DOI: 10.1007/978-3-319-57959-7

Wang, G., Zhao, Y., Huang, J., & Wu, Y. (2017). An effective approach to controller placement in software defined wide area networks. *IEEE Transactions on Network and Service Management*, *15*(1), 344–355. DOI: 10.1109/TNSM.2017.2785660

Compilation of References

ABB. (2021). ABB's Marine Advisory System: Driving sustainable shipping. https://global.abb

ABB. (2021). Marine Advisory System: Emissions monitoring for sustainable shipping. ABB Marine & Ports. https://new.abb.com/marine

Abdalla, A., Mahmoud, M., & Rizk, M. (2022). UAV anomaly detection using stacked autoencoders with temporal filtering. *Journal of Intelligent & Robotic Systems*, *104*(2), 33–45. DOI: 10.1007/s10846-022-01549-1

Abi-Zeid, I., Morin, M., & Nilo, O. (2019, May). Decision Support for Planning Maritime Search and Rescue Operations in Canada. In *ICEIS (1)* (pp. 328-339). DOI: 10.5220/0007730303280339

Acciaro, M., Ghiara, H., & Cusano, M. I. (2018). Energy management in seaports: A new role for port authorities. *Energy Policy*, *71*, 4–12. DOI: 10.1016/j.enpol.2014.04.013

Acciaro, M., & Sys, C. (2020). Innovation in the maritime sector: Aligning strategy with outcomes. *Maritime Policy & Management*, *47*(8), 1045–1063. DOI: 10.1080/03088839.2020.1737335

Ahmed, I., Jun, M., & Ding, Y. (2022). A spatio-temporal track association algorithm based on marine vessel automatic identification system data. *IEEE Transactions on Intelligent Transportation Systems*, *23*(11), 20783–20797. DOI: 10.1109/TITS.2022.3187714

Ahmed, Y. A., Theotokatos, G., Maslov, I., Wennersberg, L. A. L., & Nesheim, D. A. (2024). Regulatory and legal frameworks recommendations for short sea shipping maritime autonomous surface ships. *Marine Policy*, *166*, 106226. DOI: 10.1016/j.marpol.2024.106226

Alcaraz, C., & Zeadally, S. (2015). Critical infrastructure protection: Requirements and challenges for the 21st century. *International Journal of Critical Infrastructure Protection*, *8*, 53–66. DOI: 10.1016/j.ijcip.2014.12.002

Al-Fuqaha, A., Guizani, M., Mohammadi, M., Aledhari, M., & Ayyash, M. (2015). Internet of Things: A survey on enabling technologies, protocols, and applications. *IEEE Communications Surveys and Tutorials*, *17*(4), 2347–2376. DOI: 10.1109/COMST.2015.2444095

Allouch, A., Cheikhrouhou, O., Koubâa, A., Khalgui, M., & Abbes, T. (2019, June). MAVSec: Securing the MAVLink protocol for ardupilot/PX4 unmanned aerial systems. *In 2019 15th International Wireless Communications & Mobile Computing Conference (IWCMC)* (pp. 621–628). IEEE. DOI: 10.1109/IWCMC.2019.8766571

Almutairi, A., & Khan, R. (2023). Image-based classical features and machine learning analysis of skin cancer instances. *Applied Sciences (Basel, Switzerland)*, *13*(13), 7712. DOI: 10.3390/app13137712

Animah, I. (2024). Application of bayesian network in the maritime industry: Comprehensive literature review. *Ocean Engineering*, *302*, 117610. DOI: 10.1016/j.oceaneng.2024.117610

An, J. (2024). Maritime logistics and digital transformation with big data: Review and research trend. *Maritime Business Review*, *9*(3), 229–242. DOI: 10.1108/MABR-10-2023-0069

Arnold, A. C., Gasparoto, H. F., Brunet, J. P., Vega, P. J. S., & Violante, P. (2025). Challenges and Innovations in Drone and Autonomous Vehicle Simulators and Controllers. *OCEANS 2025 Brest*, 1-10.

Asahi, S. (2024). Economic Ripple Effects of Choice Factors and Disaster-Related Investments at Nagoya Port. In *Econometric Analysis of Regional Economy* (pp. 185–230). Springer Nature Singapore. DOI: 10.1007/978-981-99-8707-8_5

Autsadee, Y., Jeevan, J., Bin Othman, M. R., & Mohd Salleh, N. H. B. (2025). Maritime Society 5.0: A global transition on human economy and civilisation for maritime sustainability. *Australian Journal of Maritime and Ocean Affairs*, *17*(1), 1–26. DOI: 10.1080/18366503.2023.2287872

Awais, C. M., Reggiannini, M., Moroni, D., & Salerno, E. (2025). A Survey on SAR ship classification using Deep Learning. *arXiv preprint arXiv:2503.11906.*

Bai, X., Zhang, Z., & Xu, X. (2025). Advanced multispectral detection of maritime targets for unmanned ocean surveillance. Ocean Engineering. https://www.sciencedirect.com/science/article/abs/pii/S0029801825008984

Bajwa, A. (2025). AI-based emergency response systems: A systematic literature review on smart infrastructure safety. *Available at SSRN* 5171521.

Bal Beşikçi, E., Kececi, T., Arslan, O., & Turan, O. (2019). An artificial neural network- based decision support system for maritime safety management. *Safety Science, 120*, 490–500. DOI: 10.1016/j.ssci.2019.07.031

Balduzzi, M., Pasta, A., & Wilhoit, K. (2014, December). A security evaluation of AIS automated identification system. In *Proceedings of the 30th annual computer security applications conference* (pp. 436-445). https://doi.org/DOI: 10.1145/2664243.2664257

Barnett, M. L., Gatfield, D., & Pekcan, C. (2006). A research agenda in maritime crew resource management. *Proceedings of the Human Factors and Ergonomics Society Annual Meeting, 50*(19), 2081–2085. https://scispace.com/papers/a-research-agenda-in-maritime-crew-resource-management-14o7j2f2uw

Benz, L., Münch, C., & Hartmann, E. (2021). Development of a search and rescue framework for maritime freight shipping in the Arctic. *Transportation Research Part A, Policy and Practice, 152*, 54–69. DOI: 10.1016/j.tra.2021.07.006

BIMCO. (2021). Cyber security guidelines for ships. https://www.bimco.org/cyber-security

Bisong, E. (2019). Google Colaboratory. In *Building Machine Learning and Deep Learning Models on Google Cloud Platform* (pp. 59–64). Apress. DOI: 10.1007/978-1-4842-4470-8_7

Bloisi, D. D., Previtali, F., Pennisi, A., Nardi, D., & Fiorini, M. (2016). Enhancing automatic maritime surveillance systems with visual information. *IEEE Transactions on Intelligent Transportation Systems, 18*(4), 824–833. DOI: 10.1109/TITS.2016.2591321

Bochkovskiy, A., Wang, C. Y., & Liao, H. Y. M. (2020). Yolov4: Optimal speed and accuracy of object detection. *arXiv preprint arXiv:2004.10934*. https://doi.org//arXiv.2004.10934DOI: 10.48550

Bond, T. C., Doherty, S. J., Fahey, D. W., Forster, P. M., Berntsen, T., DeAngelo, B. J., & Zender, C. S. (2013). Bounding the role of black carbon in the climate system: A scientific assessment. *Journal of Geophysical Research, D, Atmospheres, 118*(11), 5380–5552. DOI: 10.1002/jgrd.50171

Brar, P. (2025). Drones Over the Strait: How Taiwan's UAV Programme is Redrawing Cross-Strait Red Lines. RSIS Commentary. https://www.rsis.edu.sg/wp-content/uploads/2025/08/CO25171.pdf

Brynolf, S., Andersson, K., & Ng, A. K. Y. (2014). Environmental assessment of marine fuels: Liquefied natural gas, liquefied biogas, methanol, and bio-methanol. *Journal of Cleaner Production*, *74*, 86–95. DOI: 10.1016/j.jclepro.2014.03.052

Brynolf, S., Fridell, E., & Andersson, K. (2022). Environmental assessment of hydrogen and ammonia as fuels for shipping. *Sustainable Energy & Fuels*, *6*(2), 300–318. DOI: 10.1039/D1SE01222F

Buczak, A. L., & Guven, E. (2015). A survey of data mining and machine learning methods for cyber security intrusion detection. *IEEE Communications Surveys and Tutorials*, *18*(2), 1153–1176. DOI: 10.1109/COMST.2015.2494502

Callcut, M., Agliozzo, J. P., Varga, L., & McMillan, L. (2021). Digital Twins in Civil Infrastructure Systems. *Sustainability (Basel)*, *13*(20), 11549. DOI: 10.3390/su132011549

Carneiro, J. P. (1988). Maritime Search and Rescue. *IETE Technical Review*, *5*(3), 111–114. DOI: 10.1080/02564602.1988.11438248

Cecchinato, N., Toma, A., Drioli, C., Oliva, G., Sechi, G., & Foresti, G. L. (2023). Secure Real-Time Multimedia Data Transmission from Low-Cost UAVs with A Lightweight AES Encryption. In *IEEE communications magazine* (Vol. 61, Number 5, pp. 160–165). IEEE. DOI: 10.1109/MCOM.001.2200611

Chalapathy, R., & Chawla, S. (2019). Deep learning for anomaly detection: A survey. *ACM Computing Surveys*, *51*(5), 1–36. DOI: 10.1145/3343440

Chander, A., & Lê, U. P. (2015). Data nationalism. *Emory Law Journal*, *64*(3), 677–739. https://ssrn.com/abstract=2577947

Cheng, S., Zhu, Y., & Wu, S. (2023). Deep learning based efficient ship detection from drone-captured images for maritime surveillance. *Ocean Engineering*, *285*, 115440. DOI: 10.1016/j.oceaneng.2023.115440

Chollet, F. (2018). *Deep learning with Python*. Manning Publications.

Colomina, I., & Molina, P. (2014). Unmanned aerial systems for photogrammetry and remote sensing: A review. *ISPRS Journal of Photogrammetry and Remote Sensing*, *92*, 79–97. DOI: 10.1016/j.isprsjprs.2014.02.013

Corbett, J. J., & Winebrake, J. J. (2008). The impacts of global maritime shipping on air pollution and climate change. *Environmental Science & Technology*, *42*(13), 4486–4493. DOI: 10.1021/es071686z

Costa, N. A., Holder, E., & MacKinnon, S. N. (2017). Implementing human-centred design in the context of a graphical user interface redesign for ship manoeuvring. *International Journal of Human-Computer Studies*, *100*, 55–65. DOI: 10.1016/j.ijhcs.2016.12.006

Cullinane, K., & Haralambides, H. (2021). Global trends in maritime and port economics: The COVID-19 pandemic and beyond. *Maritime Economics & Logistics*, *23*(3), 369–380. DOI: 10.1057/s41278-021-00196-5

De Beukelaer, C. (2024). Shipping the future: Climate ethics for maritime transport. *ephemera: theory & politics in organization, 24*(1).

Debelić, B. (2024). Economic Governance in Maritime Affairs. *Pomorski zbornik, 64*(1), 135-156.

Diniz, N. V., Cunha, D. R., de Santana Porte, M., Oliveira, C. B. M., & de Freitas Fernandes, F. (2024). A bibliometric analysis of sustainable development goals in the maritime industry and port sector. *Regional Studies in Marine Science*, *69*, 103319. DOI: 10.1016/j.rsma.2023.103319

DNV. (2021). *Maritime autonomous surface ships: Regulatory framework and challenges* [Technical and regulatory news]. DNV. Retrieved from https://www.scribd.com/document/516077075

DNV. (2021). Yara Birkeland: The world's first autonomous and electric container ship. https://www.dnv.com

Dominioni, G. (2024). Carbon pricing for international shipping, equity, and WTO law. *Review of European, Comparative & International Environmental Law*, *33*(1), 19–30. DOI: 10.1111/reel.12540

Duan, G. J., & Zhang, P. F. (2014). Research on application of UAV for maritime supervision. *Journal of Shipping and Ocean Engineering*, *4*, 322–326.

Durlik, I., Miller, T., Kostecka, E., & Tuński, T. (2024). Artificial intelligence in maritime transportation: A comprehensive review of safety and risk management applications. *Applied Sciences (Basel, Switzerland)*, *14*(18), 8420. DOI: 10.3390/app14188420

Elston, J., Pinto, H., & Nogueira, C. (2024). Tides of Change for a Sustainable Blue Economy: A Systematic Literature Review of Innovation in Maritime Activities. *Sustainability (Basel)*, *16*(24), 11141. DOI: 10.3390/su162411141

EMSA. (2022). Study on electrical energy storage systems in maritime transport. European Maritime Safety Agency. https://www.emsa.europa.eu

Eren, V., & Duman, H. (2025). Artificial Intelligence Support In Disaster Management. *Kamu Yönetimi ve Teknoloji Dergisi*, *7*(1), 13–36. DOI: 10.58307/kaytek.1580460

European Commission. (2020). *Europe's digital strategy: Shaping Europe's digital future.* European Commission. (2020). *Europe's digital strategy: Shaping Europe's digital future.* https://digital-strategy.ec.europa.eu/en?utm_source=chatgpt.com

European Commission. (2023). EU Emissions Trading System (EU ETS): Maritime sector inclusion. https://climate.ec.europa.eu

Eyring, V., Köhler, H. W., van Aardenne, J., & Lauer, A. (2010). Emissions from international shipping: 1. The last 50 years. *Journal of Geophysical Research, D, Atmospheres*, *115*(D12). Advance online publication. DOI: 10.1029/2009JD012396

Farah, M. B., Ahmed, Y., Mahmoud, H., Shah, S. A., Al-Kadri, M. O., Taramonli, S., Bellekens, X., Abozariba, R., Idrissi, M., & Aneiba, A. (2024). A survey on blockchain technology in the maritime industry: Challenges and future perspectives. *Future Generation Computer Systems*, *157*, 618–637. DOI: 10.1016/j.future.2024.03.046

Ficco, M., Palmiero, R., Rak, M., & Granata, D. (2022). *MAVLink protocol for unmanned aerial vehicle: Vulnerabilities analysis. 2022 IEEE International Conference on Dependable, Autonomic and Secure Computing (DASC), Pervasive Intelligence and Computing (PiCom), Cloud and Big Data Computing (CBDCom), and Cyber Science and Technology Congress (CyberSciTech)*, 1–6. DOI: 10.1109/DASC/PiCom/CBDCom/Cy55231.2022.9927895

Frederiksen, M. H., & Knudsen, M. P. (2018). Drones for offshore and maritime missions: Opportunities and barriers.

Gans, N. R., & Rogers, J. G.III. (2021). Cooperative multirobot systems for military applications. *Current Robotics Reports*, *2*(1), 105–111. DOI: 10.1007/s43154-020-00039-w

Garay-Rondero, C. L., & Issa-Zadeh, S. B. (2025). Integrating sustainability thinking in maritime education. *Cogent Education*, *12*(1), 2479205. DOI: 10.1080/2331186X.2025.2479205

Ghazlane, Y., Hajar, B., Hajar, E. M., Saida, H., & Mourad, E. Y. (2025). Artificial Intelligence in Drone and Anti-drone Systems: Cutting-Edge Advances and Applications. In *International Conference on Electrical Systems and Smart Technologies* (pp. 167-174). Springer, Cham. DOI: 10.1007/978-3-031-86705-7_16

Globisch, T., Umlauft, A., Lüdtke, S., & Von Lukas, U. F. (2024, September). sMarT RCI-Unsupervised Double K-Means Ensemble Clustering Tool for Remotely Operated Vehicle Camera Images. In *OCEANS 2024-Halifax* (pp. 1-10). IEEE. https://doi.org/DOI: 10.1109/OCEANS55160.2024.10754492

González Fuster, G. (2019). *The emergence of personal data protection as a fundamental right of the EU*. Springer., DOI: 10.1093/idpl/ipu034

Gozalan, A., John, O., Lubcke, T., Maier, A., Reimann, M., Richter, J. G., & Zverev, I. (2020). Assisting maritime search and rescue (SAR) personnel with AI-based speech recognition and smart direction finding. *Journal of Marine Science and Engineering*, *8*(10), 818. DOI: 10.3390/jmse8100818

Grand View Research. (2024). *Maritime artificial intelligence market: Industry report, 2030*. Grand View Research. https://www.grandviewresearch.com/industry analysis/maritime-artificial-intelligence-market-report

Grech, M. R., Horberry, T., & Koester, T. (2019). *Human factors in the maritime domain*. CRC Press., DOI: 10.1201/9780429355417

Gupta, A., Bhatnagar, A., & Mehta, A. (2021). Application of drones in maritime industry (firefighting). *Bulletin of Marine Science and Technology*, *14*, 59–69.

Hale, J., Long, S., Corns, S., & Shoberg, T. (2019). Flood management deep learning model inputs: A review of necessary data and predictive tools. *Proceedings of the International Annual Conference of the American Society for Engineering Management (ASEM)*, 1–9. DOI: 10.5465/AMBPP.2019.15761

Hall, D., & Llinas, J. (2001). *Multisensor data fusion*. CRC press., DOI: 10.1201/9781420038545

Hamza, M. A., Mohsin, M., Khalil, M., & Kazam Abbas Kazmi, S. M. (2024). MAVLink Protocol: A Survey of Security Threats and Countermeasures. *2024 4th International Conference on Digital Futures and Transformative Technologies (ICoDT2)*, 1–8. DOI: 10.1109/ICoDT262145.2024.10740195

Hashesh, A. O., Hashima, S., Zaki, R. M., Fouda, M. M., Hatano, K., & Eldien, A. S. T. (2022). AI-enabled UAV communications: Challenges and future directions. *IEEE Access : Practical Innovations, Open Solutions*, *10*, 92048–92066. DOI: 10.1109/ACCESS.2022.3202956

Haugen, S., & Kristiansen, S. (2022). *Maritime transportation: Safety management and risk analysis* (2nd ed.). Routledge., DOI: 10.4324/9781003055464

Hebbar, A. A. (2024). Implementation of IMO instruments from a flag state perspective. *The Elgar Companion to the Law and Practice of the International Maritime Organization*, 304-329.

Heilig, L., Lalla-Ruiz, E., & Voß, S. (2017). Digital transformation in maritime ports: Analysis and a game theoretic framework. *NETNOMICS: Economic Research and Electronic Networking*, *18*(2–3), 227–254. DOI: 10.1007/s11066-017-9122-x

Hetherington, C., Flin, R., & Mearns, K. (2006). Safety in shipping: The human element. *Journal of Safety Research*, *37*(4), 401–411. DOI: 10.1016/j.jsr.2006.04.007 PMID: 17046789

Hofmann, E., & Osterwalder, F. (2017). Third-party logistics providers in the digital age: Towards a new competitive arena? *Logistics (Basel)*, *1*(2), 1–28. DOI: 10.3390/logistics1020009

IMO. (2018). Initial IMO Strategy on Reduction of GHG Emissions from Ships. https://www.imo.org

IMO. (2020). Fourth, IMO, GHG Study 2020. International Maritime Organization. https://www.imo.org

IMO. (2020). MARPOL Annex VI: Prevention of air pollution from ships. https://www.imo.org

IMO. (2021). Carbon Intensity Indicator (CII) and Energy Efficiency Existing Ship Index (EEXI). https://www.imo.org

IMO. (2021). Regulatory scoping exercise for the use of Maritime Autonomous Surface Ships (MASS). https://www.imo.org

Inal, O. B. (2024). Decarbonization of shipping: Hydrogen and fuel cells legislation in the maritime industry. *Brodogradnja*, *75*(2), 1–13. DOI: 10.21278/brod75205

International Journal of Trend in Scientific Research and Development. (2024). Drones in Maritime Industry. IJTSRD, 8(4), 520–528. https://www.ijtsrd.com/papers/ijtsrd97093.pdf

International Maritime Organization (IMO). (2021). *Guidelines on maritime cyber risk management.* https://www.imo.org

International Maritime Organization. (2021). World maritime theme: Seafarers at the core of shipping's future. IMO. https://www.imo.org

International Maritime Organization. (2025, March 18). *IMO to develop a global strategy for maritime digitalization*. https://www.imo.org/en/MediaCentre/PressBriefings/pages/IMO-global-strategy maritime-digitalization.aspx

Islam, M. S. (2024). Legal implications of emerging technologies in maritime pollution monitoring and management. *Journal of the Indian Ocean Region*, *20*(1), 54–75. DOI: 10.1080/19480881.2024.2387819

Islam, S. R., Kwak, D., Kabir, M. H., Hossain, M., & Kwak, K. S. (2015). The Internet of Things for health care: A comprehensive survey. *IEEE Access : Practical Innovations, Open Solutions*, *3*, 678–708. DOI: 10.1109/ACCESS.2015.2437951

IWSA. (2021). Wind propulsion technologies and systems. International Windship Association. https://wind-ship.org

Ji, J., Chi, Y., & Yin, X. (2024). Research on the driving effect of marine economy on the high-quality development of regional economy–Evidence from China's coastal areas. *Regional Studies in Marine Science*, *74*, 103550. DOI: 10.1016/j.rsma.2024.103550

Jin, Z., He, T., Qiao, L., Duan, J., Shi, X., Yan, B., & Guo, C. (2025). MES-YOLO: An efficient lightweight maritime search and rescue object detection algorithm with improved feature fusion pyramid network. *Journal of Visual Communication and Image Representation*, *109*, 104453. DOI: 10.1016/j.jvcir.2025.104453

Kaber, D. B., & Endsley, M. R. (2004). The effects of level of automation and adaptive automation on human performance, situation awareness, and workload in a dynamic control task. *Theoretical Issues in Ergonomics Science*, *5*(2), 113–153. DOI: 10.1080/1463922021000054335

Karakas, S., Acar, A. Z., & Kirmizi, M. (2024). Maritime Sustainability: Navigating Complex Challenges and Ecological Footprints. In *Sustainable Development Seen Through the Lenses of Ethnoeconomics and the Circular Economy* (pp. 19–47). Springer Nature Switzerland. DOI: 10.1007/978-3-031-72676-7_2

Kilic, K. I., Maity, S., Sung, I., & Nielsen, P. (2025). Challenges and AI-driven solutions in maritime search and rescue planning: A comprehensive literature review. *Marine Policy*, *178*, 106692. DOI: 10.1016/j.marpol.2025.106692

Klcin, N. (2021). Maritime autonomous vehicles and international laws on boat migration: Lessons from the use of drones in the Mediterranean. *Marine Policy*, *127*, 104447. DOI: 10.1016/j.marpol.2021.104447

Konert, A., & Balcerzak, T. (2021). Military autonomous drones (UAVs)-from fantasy to reality. Legal and Ethical implications. *Transportation Research Procedia*, *59*, 292–299. DOI: 10.1016/j.trpro.2021.11.121

Kong, S. G., Heo, J., Abidi, B. R., Paik, J., & Abidi, M. A. (2005). Recent advances in visual and infrared face recognition—A review. *Computer Vision and Image Understanding*, *97*(1), 103–135. DOI: 10.1016/j.cviu.2004.04.001

Koo, K. Y., Rødseth, Ø. J., Lislebø, G., & Ulvensøen, J. H. (2024, October). Harmonizing Maritime Innovation: Enhancing International and National Standardization in Intelligent Ship Transport Systems. [). IOP Publishing.]. *Journal of Physics: Conference Series*, *2867*(1), 012023. DOI: 10.1088/1742-6596/2867/1/012023

Kraska, J. (2010). The law of unmanned naval systems in war and peace. *Journal of Ocean Technology*.

Krystosik-Gromadzińska, A. (2021). The use of drones in maritime sector–areas and benefits. *Polish Maritime Research*, *28*, 85–91.

Kumar, S., & Sharma, N. (2022). Emerging military applications of free space optical communication technology: A detailed review. [). IOP Publishing.]. *Journal of Physics: Conference Series*, *2161*(1), 012011. DOI: 10.1088/1742-6596/2161/1/012011

Kusic, K., Schuman, R., & Ivanjko, E. (2023). A digital twin in transportation: Real-time synergy of traffic data streams and simulation for virtualizing motorway dynamics. *Advanced Engineering Informatics*, *55*, 101858. Advance online publication. DOI: 10.1016/j.aei.2022.101858

Kusumawati, E. D., & Punomo, B. (2024). Technological Transformation and Innovation Strategy as a Key Pillar in Improving Sustainability Efficiency In The Maritime Industry. *IITE Proceeding: International Inovation Technology Proceeding, 2*(1), 65-76.

Lam, J. S. L., & Notteboom, T. (2014). The greening of ports: A comparison of port management tools used by leading ports in Asia and Europe. *Transport Reviews*, *34*(2), 169–189. DOI: 10.1080/01441647.2014.891162

Leman, J., & Nevalainen, P. (2025). The Future of Unmanned Naval Systems: An Overview Of Efficiency Enhancements and Warfare Upgrades for Maritime Drones. *Robotics and Autonomous Systems*.

Levander, O. (2017). Autonomous ships on the high seas. *IEEE Spectrum*, *54*(2), 26–31. DOI: 10.1109/MSPEC.2017.7849225

Liang, M., Weng, L., Gao, R., Li, Y., & Du, L. (2024). Unsupervised maritime anomaly detection for intelligent situational awareness using AIS data. *Knowledge-Based Systems*, *284*, 111313. DOI: 10.1016/j.knosys.2023.111313

Liu, J., Liao, R., Dong, F., Huang, C., Li, H., Liu, J., & Zhao, T. (2024). Low-carbon technology selection and carbon reduction potential assessment in the shipbuilding industry with dynamically changing grid emission factors. *Journal of Cleaner Production*, *441*, 140707. DOI: 10.1016/j.jclepro.2024.140707

Liu, R. W., Liang, M., Nie, J., Lim, W. Y. B., Zhang, Y., & Guizani, M. (2022). Deep learning-powered vessel trajectory prediction for improving smart traffic services in maritime Internet of Things. *IEEE Transactions on Network Science and Engineering*, *9*(5), 3080–3094. DOI: 10.1109/TNSE.2022.3140529

Liu, R. W., Liang, M., Nie, J., Yuan, Y., Xiong, Z., Yu, H., & Guizani, N. (2022). STMGCN: Mobile edge computing-empowered vessel trajectory prediction using spatio-temporal multigraph convolutional network. *IEEE Transactions on Industrial Informatics*, *18*(11), 7977–7987. DOI: 10.1109/TII.2022.3165886

Liu, Z., Schindler, S., & Liu, W. (2020). Demystifying Chinese overseas investment in infrastructure: Port development, the Belt and Road Initiative and regional development. *Journal of Transport Geography*, *87*, 102812. Advance online publication. DOI: 10.1016/j.jtrangeo.2020.102812

Li, Y., Yuan, H., Wang, Y., & Xiao, C. (2022). GGT-YOLO: A novel object detection algorithm for drone-based maritime cruising. *Drones (Basel)*, *6*(11), 335. DOI: 10.3390/drones6110335

Lu, Y., Liu, C., Wang, K. I.-K., Huang, H., Xu, X., & Zhang, W. (2021). Industrial Internet of Things (IIoT). A smart manufacturing perspective. *IEEE Transactions on Industrial Informatics*, *17*(3), 2390–2402. DOI: 10.1109/TII.2019.2942495

Lu, Y., Liu, C., & Xu, X. (2021). Digital twin-driven smart maintenance: Systems, technologies, and applications. *Journal of Manufacturing Systems*, *58*, 346–361. DOI: 10.1016/j.jmsy.2020.06.006

Lynch, K. M., Banks, V. A., Roberts, A. P., Radcliffe, S., & Plant, K. L. (2024). What factors may influence decision-making in the operation of Maritime autonomous surface ships? A systematic review. *Theoretical Issues in Ergonomics Science*, *25*(1), 98–142. DOI: 10.1080/1463922X.2022.2152900

Madsen, A. N., & Kim, T. E. (2024). A state-of-the-art review of AI decision transparency for autonomous shipping. *Journal of International Maritime Safety, Environmental Affairs, and Shipping*, *8*(1-2), 2336751. DOI: 10.1080/25725084.2024.2336751

Mallam, S. C., Nazir, S., & Sharma, A. (2020). The human element in future maritime operations—Perceived impact of autonomous shipping. *Ergonomics*, *63*(3), 334–345. DOI: 10.1080/00140139.2019.1659998

Mansouri, Y., Toosi, A. N., & Buya, R. (2017). Data Storage Management in Cloud Environments: Taxonomy, Survey, and Future Directions. *ACM Computing Surveys*, *50*(6), 1–51. Advance online publication. DOI: 10.1145/3136623

Manyika, J., Lund, S., Bughin, J., Woetzel, J., Stamenov, K., & Dhingra, D. (2021). *Digital globalization: The new era of global flows.* McKinsey Global Institute. Retrieved from: https://www.mckinsey.com/capabilities/mckinsey-digital/our-insights/digital-globalization-the-new-era-of-global-flows

Mao, H., Alizadeh, M., Menache, I., & Kandula, S. (2016). Resource management with deep reinforcement learning. In *Proceedings of the 15th ACM Workshop on Hot Topics in Networks* (pp. 50–56). ACM. https://doi.org/DOI: 10.1145/3005745.3005750

Martinez-Esteso, J. P., Castellanos, F. J., Calvo-Zaragoza, J., & Gallego, A. J. (2024). Maritime Search and Rescue Missions with Aerial Images: A Survey. *arXiv preprint arXiv:2411.07649.*

Martinez-Esteso, J. P., Castellanos, F. J., Calvo-Zaragoza, J., & Gallego, A. J. (2025). Maritime search and rescue missions with aerial images: A survey. *Computer Science Review*, *57*, 100736. DOI: 10.1016/j.cosrev.2025.100736

Mascaro, S., Korb, K.B., & Nicholson, A.E. (2010). Learning Abnormal Vessel Behaviour from AIS Data with Bayesian Networks at Two Time Scales.

Mascaro, S., Nicholso, A. E., & Korb, K. B. (2014). Anomaly detection in vessel tracks using Bayesian networks. *International Journal of Approximate Reasoning*, *55*(1), 84–98. DOI: 10.1016/j.ijar.2013.03.012

Mohammed, M. A., & Ahmed, A. (2025). Artificial intelligence applications in cancer diagnostic devices: A focused review. *Medical Oncology Journal*. Advance online publication. DOI: 10.61706/MECOJ160160

Mukherjee, T. (2018). *The role of artificial intelligence in naval operations*. ORF Observer Research Foundation.

Nair, C. D. J. N. (Ed.). (2023). *Emerging Defence, Maritime and Aerospace Technologies by DRaS*. Highlyy Publishing LLP.

Nikolic, D., Stojkovic, N., Puzovic, S., Popovic, Z., Stojiljkovic, N., Grbic, N., & Orlic, V. D. (2023). Increasing Maritime Safety and Security in the Off-Shore Activities with HFSWRs as Primary Sensors for Risk Assessment. *Journal of Marine Science and Engineering*, *11*(6), 1167. https://www.mdpi.com/2077-1312/11/6/1167. DOI: 10.3390/jmse11061167

Nomikos, N., Gkonis, P. K., Bithas, P. S., & Trakadas, P. (2022). A survey on UAV-aided maritime communications: Deployment considerations, applications, and future challenges. *IEEE Open Journal of the Communications Society*, *4*, 56–78. DOI: 10.1109/OJCOMS.2022.3225590

Notteboom, T., & Haralambides, H. (2020). Port management and governance in a post-COVID-19 era: Quo vadis? *Maritime Economics & Logistics*, *22*(3), 329–352. DOI: 10.1057/s41278-020-00162-7

Okumus, D., Gunbeyaz, S. A., Kurt, R. E., & Turan, O. (2024). An approach to advance circular practices in the maritime industry through a database as a bridging solution. *Sustainability (Basel)*, *16*(1), 453. DOI: 10.3390/su16010453

Olaniyi, E. O., Solarte-Vasquez, M. C., & Inkinen, T. (2024). Smart regulations in maritime governance: Efficacy, gaps, and stakeholder perspectives. *Marine Pollution Bulletin*, *202*, 116341. DOI: 10.1016/j.marpolbul.2024.116341 PMID: 38608427

Pacific Data Integrators. (n.d.). *PDI customer care: Revolutionize financial services with generative AI*. Pacific Data Integrators. https://www.pacificdataintegrators.com/industry solution/financial-services-with- generative-ai

Palmer-Abbs, M., Deshpande, P., & Karl, C. W. (2024). Enabling transition thinking on complex issues (wicked problems): A framework for future circular economic transitions of plastic management in the Norwegian fisheries and aquaculture sectors. *Journal of Cleaner Production*, *449*, 141420. DOI: 10.1016/j.jclepro.2024.141420

Pal, O. K., Shovon, M. D., Mridha, M. F., & Shin, J. (2024). In-depth review of AI-enabled unmanned aerial vehicles: Trends, vision, and challenges. *Discover Artificial Intelligence*, *4*(1), 1–24. DOI: 10.1007/s44163-024-00209-1

Panić, I., Ćelić, J., Bistrović, M., & Škrobonja, A. (2021). Drone as a part of maritime search and rescue operations. *Technologies, Techniques and Applications Across PNT*, 63.

Panico, A., Fragonara, L. Z., & Al-Rubaye, S. (2020). Adaptive detection tracking system for autonomous UAV maritime patrolling. In 2020 IEEE 7th International Workshop on Metrology for AeroSpace (MetroAeroSpace), 539–544. https://dspace.lib.cranfield.ac.uk/items/f5feff39-3e8c-44bd-bfdd-c6d9478911b9

Papyan, N., Kulhandjian, M., Kulhandjian, H., & Aslanyan, L. (2024). Ai-based drone assisted human rescue in disaster environments: Challenges and opportunities. *Pattern Recognition and Image Analysis, 34*(1), 169–186. DOI: 10.1134/S1054661824010152

Pareek, P., Hajra, R., & Mohapatra, H. (2025). Empowering Public Officials by Bridging the Digital Divide with Smart Technology for Public Services. *Digital Competency Development for Public Officials: Adapting New Technologies in Public Services*, 419-452.

Pedregosa, F., Varoquaux, G., & Gramfort, A.. (2011). Scikit-learn: Machine learning in Python. *Journal of Machine Learning Research, 12*, 2825–2830.

Perera, M. V., Bandara, W. G. C., Valanarasu, J. M. J., & Patel, V. M. (2022, July). Transformer-based SAR image despeckling. In *IGARSS 2022-2022 IEEE International Geoscience and Remote Sensing Symposium* (pp. 751-754). IEEE. DOI: 10.1109/IGARSS46834.2022.9884596

Pettersson, E. (2020). Design of a drone system for maritime search and rescue missions.

Porathe, T., Ehlers, S., & Jensen, M. (2018). Maritime Autonomous Surface Ships (MASS): State-of-the-art autonomous control. Proceedings of the 5th International Conference on Maritime Science and Technology. Port of Rotterdam. (2023). Digital and automated port operations. https://www.portofrotterdam.com

Potamos, G., Stavrou, E., & Stavrou, S. (2024). Enhancing maritime cybersecurity through operational technology sensor data fusion: A comprehensive survey and analysis. *Sensors (Basel), 24*(11), 3458. DOI: 10.3390/s24113458 PMID: 38894249

Prasad, D. K., Rajan, D., Rachmawati, L., Rajabally, E., & Quek, C. (2017). Video processing from electro-optical sensors for object detection and tracking in a maritime environment: A survey. *IEEE Transactions on Intelligent Transportation Systems, 18*(8), 1993–2016. DOI: 10.1109/TITS.2016.2634580

Psaraftis, H. N. (2019). Sustainable shipping: A cross-disciplinary view. Springer. Smith, T. W. P., Jalkanen, J. P., Anderson, B. A., Corbett, J. J., & Faber, J. (2016). CO_2 emissions from international shipping: Possible reduction targets and their associated pathways. IMO MEPC 68/INF.24. UNCTAD. (2021). Review of Maritime Transport 2021. United Nations Conference on Trade and Development. https://unctad.org

Quero, C. O., & Martinez-Carranza, J. (2025). Unmanned aerial systems in search and rescue: A global perspective on current challenges and future applications. *International Journal of Disaster Risk Reduction*, *118*, 105199. DOI: 10.1016/j.ijdrr.2025.105199

Ramesh, P. S., & Jeyan, J. M. L. (2020, December). Comparative analysis of the impact of operating parameters on military and civil applications of mini unmanned aerial vehicle (UAV). In. AIP conference proceedings: Vol. 2311. *No. 1* (p. 030034). AIP Publishing LLC. DOI: 10.1063/5.0033989

Reddy, A. S., Ashraf, M., Manideepak, T., & Mohapatra, H. (2025). A comprehensive analysis of governance and regulatory frameworks for cloud-based initiatives for smart cities. In *Organizational sociology in the digital age* (pp. 127–146). IGI Global Scientific Publishing. DOI: 10.4018/979-8-3693-7398-9.ch008

Rejeb, A., Rejeb, K., Simske, S., & Treiblmaier, H. (2021). Humanitarian drones: A review and research agenda. *Internet of Things : Engineering Cyber Physical Human Systems*, *16*, 100434. DOI: 10.1016/j.iot.2021.100434

Rekavandi, A. M., Xu, L., Boussaid, F., Seghouane, A. K., Hoefs, S., & Bennamoun, M. (2025). A Guide to Image-and Video-Based Small Object Detection Using Deep Learning: Case Study of Maritime Surveillance. *IEEE Transactions on Intelligent Transportation Systems*, *26*(3), 2851–2879. DOI: 10.1109/TITS.2025.3530678

Reports Insights Consulting Pvt Ltd. (2025). Taiwan maritime drone market analysis. https://www.linkedin.com/pulse/taiwan-maritime-drone-market-analysis-key-developments-o6kgf

Riadh, M. (2024). Evaluating the environmental and economic impacts of MARPOL compliance in the global shipping industry. *Collaborate Engineering Daily Book Series*, *2*(2), 98–104. DOI: 10.62012/collaborate.v2i2.70

Riveiro, M., Falkman, G., & Ziemke, T. (2008). Visual analytics for the detection of anomalous maritime behavior. In *Proceedings of the International Conference on Information Visualisation* (pp. 273–279). IEEE. https://doi.org/DOI: 10.1109/IV.2008.25

Riveiro, M., Pallotta, G., & Vespe, M. (2018). Maritime anomaly detection: A review. *Wiley Interdisciplinary Reviews. Data Mining and Knowledge Discovery*, *8*(5), e1266. DOI: 10.1002/widm.1266

Rodrigue, J. P. (2020). *The geography of transport systems* (5th ed.). Routledge., DOI: 10.4324/9780429346323

Rodseth, O. J., & Burmeister, H.-C. (2015). *Risk assessment for autonomous ships. Transactions on Maritime Science, 4*(1), 31–40. Retrieved from https://www.transnav.eu/Article_Risk_Assessment_for_an_Unmanned_R%C3%B8dseth%2C35%2C593.html

Ruiz Estrada, M. A. (2024). A New Advanced Military Drones Systems: Prototypes Modeling. *Available at SSRN* 4912793. DOI: 10.2139/ssrn.4912793

Samanth, S., K v, P., & Balachandra, M. (2022). Security in the Internet of Drones: A comprehensive review. *Cogent Engineering*, *9*(1), 2029080. DOI: 10.1080/23311916.2022.2029080

Sameera Sulakshana, B. K. (2024). *New technologies to enhance GMDSS and reduce workload on deck officers* [PDF]. CORE. https://core.ac.uk/download/638723085.pdf

Satyanarayanan, M. (2017). The emergence of edge computing. *Computer*, *50*(1), 30–39. DOI: 10.1109/MC.2017.9

SaveOurBlueOcean. (n.d.). *Encouraging the adoption of sustainable shipping practices to minimize marine pollution*. Retrieved August 2025, from https://www.saveourblueocean.shop/post/encouraging-the-adoption-of-sustainable shipping-practices-to-minimize-marine-pollution

Schill, N., & Browning, M. C. K. (2015). The impact of satellite AIS to the environmental challenges of modern shipping. In *Maritime Informatics* (pp. 1–14). Institution of Engineering and Technology. DOI: 10.1049/PBTR001E_ch13

Sepehri, A., Kirichek, A., van den Heuvel, M., & van Koningsveld, M. (2024). Smart, sustainable, and circular port maintenance: A comprehensive framework and multi-stakeholder approach. *Journal of Environmental Management*, *370*, 122625. DOI: 10.1016/j.jenvman.2024.122625 PMID: 39316882

Sheikh, W., Islam, T. R., & Islam, H. (2024). Analyzing the impact of organizational justice on seafarers' commitment: Job satisfaction as a key mediator. *Australian Journal of Maritime and Ocean Affairs*, •••, 1–31. DOI: 10.1080/18366503.2024.2403741

Shi, W., Cao, J., Zhang, Q., Li, Y., & Xu, L. (2016). Edge computing: Vision and challenges. *IEEE Internet of Things Journal*, *3*(5), 637–646. DOI: 10.1109/JIOT.2016.2579198

Singapore, M. P. A. (2023). Smart port initiatives. Maritime and Port Authority of Singapore. https://www.mpa.gov.sg

Singh, A. (2016). Unmanned and autonomous vehicles and future maritime operations in littoral asia. *ORF Special Report, 28*.

Sneha, S., & Surekha, P. (2025). Maritime human drone teaming for search and rescue operations. IEEE Xplore. https://ieeexplore.ieee.org/document/10531426/

Sommer, R., & Paxson, V. (2010). Outside the closed world: On using machine learning for network intrusion detection. *2010 IEEE Symposium on Security and Privacy*, 305–316. https://doi.org/DOI: 10.1109/SP.2010.25

Song, Z. Y., Chhetri, P., Ye, G., & Tae-Woo Lee, P. (2025). Green maritime logistics coalition by green shipping corridors: A new paradigm for the decarbonisation of the maritime industry. *International Journal of Logistics*, *28*(4), 363–379. DOI: 10.1080/13675567.2023.2256243

Stanton, N. A., Salmon, P. M., Rafferty, L., Walker, G. H., Baber, C., & Jenkins, D. P. (2017). *Human factors methods: A practical guide for engineering and design* (2nd ed.). CRC Press., DOI: 10.1201/9781315587394

Stopford, M. (2009). *Maritime economics* (3rd ed.). Routledge., DOI: 10.4324/9780203891742

Sumartono, E. (2024). The Life of Seafarers in the Digital Age: An Analysis of the Philosophy of Ethics and Morality in the Use of Technology. *Side. Scientific Development Journal*, *1*(3), 78–87.

Sun, X., Guo, Y., & Huang, Y. (2020). Anomaly detection in UAV telemetry using GRU-based deep learning. *Journal of Aerospace Information Systems*, *17*(11), 623–634.

Suresh Babu, C. V., Suruthi, G., Indhumathi, C., & Sushruth, S. (2024). Integrated approach for network intrusion detection: Combining KNN algorithm, Karhunen-Loève transform, and genetic algorithm optimization. In Metaheuristic and machine learning optimization strategies for complex systems (p. 21). DOI: 10.4018/979-8-3693-7842-7.ch002

Suresh Babu, C. V., Andrew Simon, P., & Barath Kumar, S. (2023). The Future of Cyber Security Starts Today, Not Tomorrow. In Shiva Darshan, S., Manoj Kumar, M., Prashanth, B., & Vishnu Srinivasa Murthy, Y. (Eds.), *Malware Analysis and Intrusion Detection in Cyber-Physical Systems* (pp. 348–375). IGI Global., DOI: 10.4018/978-1-6684-8666-5.ch016

Szegedy, C., Vanhouckc, V., Ioffe, S., Shlens, J., & Wojna, Z. (2016). Rethinking the inception architecture for computer vision. In *Proceedings of the IEEE conference on computer vision and pattern recognition* (pp. 2818-2826). **https://doi.org/**DOI: 10.1109/CVPR.2016.308

Taddeo, M., & Floridi, L. (2018). How AI can be a force for good. *Science, 361*(6404), 751–752. DOI: 10.1126/science.aat5991 PMID: 30139858

Talpur, K., Hasan, R., Gocer, I., Ahmad, S., & Bhuiyan, Z. (2025). AI in Maritime Security: Applications, Challenges, Future Directions, and Key Data Sources. *Information (Basel), 16*(8), 658. DOI: 10.3390/info16080658

Tam, K., & Jones, K. (2019). Cyber-risk assessment for autonomous ships. *Reliability Engineering & System Safety, 195*, 106698. DOI: 10.1109/CyberSecPODS.2018.8560690

Taylor, A. B. (2021). *Counter-Unmanned Aerial Vehicles Study: Shipboard Laser Weapon System Engagement Strategies for Countering Drone Swarm Threats in The Maritime Environment* (Doctoral dissertation, Monterey, CA; Naval Postgraduate School).

Tjia, M., Kim, A., Wijaya, E. W., Tefara, H., & Zhu, K. (2024). *Enhancing Robustness of Human Detection Algorithms in Maritime SAR through Augmented Aerial Images to Simulate Weather Conditions.* arXiv preprint arXiv:2408.13766.

Tu, E., Zhang, G., Mao, S., Rachmawati, L., & Huang, G. (2020). Modeling Historical AIS Data For Vessel Path Prediction: A Comprehensive Treatment. *ArXiv*. https://arxiv.org/abs/2001.01592

Tu, E., Zhang, G., Rachmawati, L., Rajabally, E., & Huang, G. B. (2017). Exploiting AIS data for intelligent maritime navigation: A comprehensive survey from data to methodology. *IEEE Transactions on Intelligent Transportation Systems, 19*(5), 1559–1582. DOI: 10.1109/TITS.2017.2724551

UNCTAD. (2023). Review of Maritime Transport 2023. https://unctad.org

United Nations Conference on Trade and Development (UNCTAD). (2023). *Review of maritime transport 2023.* United Nations. https://unctad.org/webflyer/review-maritime-transport-2023

Vallero, G., Renga, D., Meo, M., & Ajmone Marsan, M. (2020). Processing ANN traffic predictions for RAN energy efficiency. In *Proceedings of the 23rd International ACM Conference on Modeling, Analysis and Simulation of Wireless and Mobile Systems* (MASWiM '20). ACM. DOI: 10.1145/3416010.3423222

Venskus, J., Treigys, P., Bernatavičienė, J., Tamulevičius, G., & Medvedev, V. (2019). Real-time maritime traffic anomaly detection based on sensors and history data embedding. *Sensors (Basel), 19*(17), 3782. DOI: 10.3390/s19173782 PMID: 31480449

Verified Market Reports. (2025). Maritime Drone Market Size and Forecast. Retrieved from https://www.verifiedmarketreports.com/product/maritime-drone-market/

Voigt, P., & Von dem Bussche, A. (2017). *The EU General Data Protection Regulation (GDPR): A practical guide*. Springer., DOI: 10.1007/978-3-319-57959-7

Voroniuk, N. (2024). Relevance of international legal and national standards of safety of navigation and the role of classification societies in the context of modern challenges and innovations in shipbuilding. *Scientific Journal of the National Academy of Internal Affairs*, *4*(29), 22–34. DOI: 10.56215/naia-herald/4.2024.22

Wang, G., Zhao, Y., Huang, J., & Wu, Y. (2017). An effective approach to controller placement in software defined wide area networks. *IEEE Transactions on Network and Service Management*, *15*(1), 344–355. DOI: 10.1109/TNSM.2017.2785660

Wang, J., Zhou, K., Xing, W., Li, H., & Yang, Z. (2023). Applications, evolutions, and challenges of drones in maritime transport. *Journal of Marine Science and Engineering*, *11*(11), 2056. DOI: 10.3390/jmse11112056

Wang, T., & He, J. (2021). Artificial intelligence applications in maritime transportation: A review and future research directions. *Ocean Engineering*, *237*, 109680. DOI: 10.1016/j.oceaneng.2021.109680

Wang, X., Guo, Y., & Gao, Y. (2024). Unmanned autonomous intelligent system in 6G non-terrestrial network. *Information (Basel)*, *15*(1), 38. DOI: 10.3390/info15010038

Wang, X., Ma, X., & Grimson, W. E. L. (2008). Unsupervised activity perception in crowded and complicated scenes using hierarchical bayesian models. *IEEE Transactions on Pattern Analysis and Machine Intelligence*, *31*(3), 539–555. DOI: 10.1109/TPAMI.2008.87 PMID: 19147880

Wang, Y., & Han, Y. (2025). Measuring maritime search and rescue (SAR) accessibility using an improved spatiotemporal two-step floating catchment area method: A case study in the South China Sea. *Journal of Transport Geography*, *124*, 104155. DOI: 10.1016/j.jtrangeo.2025.104155

Wärtsilä. (2022). Fleet operations solution for sustainable shipping. Wärtsilä Voyage. https://www.wartsila.com

Winnes, H., Styhre, L., & Fridell, E. (2015). Reducing GHG emissions from ships in port areas. *Research in Transportation Business & Management*, *17*, 73–82. DOI: 10.1016/j.rtbm.2015.10.008

Winward, M. (2021). Intelligence capacity and mass violence: Evidence from Indonesia. *Comparative Political Studies*, *54*(3-4), 553–584.

Wu, W., Liu, A., Hu, J., Mo, Y., Xiang, S., Duan, P., & Liang, Q. (2024). EUAVDet: An efficient and lightweight object detector for UAV aerial images with an edge-based computing platform. *Drones (Basel)*, *8*(6), 261. DOI: 10.3390/drones8060261

Xia, W., Xing, K., & Guo, Z. (2022). Target tracking method of shipborne navigation radar. In J. C. Hung, N. Y. Yen, & J. W. Chang (Eds.), *Frontier Computing* (Lecture Notes in Electrical Engineering, Vol. 827) (pp. …). Springer Singapore. DOI: 10.1007/978-981-16-8052-6_262

Yahuza, M., Idris, M. Y. I., Ahmedy, I. B., Wahab, A. W. A., Nandy, T., Noor, N. M., & Bala, A. (2021). Internet of Drones security and privacy issues: Taxonomy and open challenges. *IEEE Access : Practical Innovations, Open Solutions*, *9*, 57243–57270. DOI: 10.1109/ACCESS.2021.3072030

Yang, C., Xu, Q., & Li, Z. (2019). Ship safety monitoring system based on the Internet of Things. *Journal of Coastal Research*, *93*(sp1), 698–702. DOI: 10.2112/SI93-094.1

Yao, H., Qin, R., & Chen, X. (2019). Unmanned aerial vehicle for remote sensing applications—A review. *Remote Sensing (Basel)*, *11*(12), 1443. DOI: 10.3390/rs11121443

Yao, K., & Zheng, Y. (2023). *Nanophotonics and machine learning: Concepts, fundamentals, and applications*. Springer., DOI: 10.1007/978-3-031-20473-9

Youk, G., & Kim, M. (2023). Transformer-based synthetic-to-measured SAR image translation via learning of representational features. *IEEE Transactions on Geoscience and Remote Sensing*, *61*, 1–18. DOI: 10.1109/TGRS.2023.3267480

Yucesoy, E., Balcik, B., & Coban, E. (2025). The role of drones in disaster response: A literature review of operations research applications. *International Transactions in Operational Research*, *32*(2), 545–589. DOI: 10.1111/itor.13484

Zaman, F. H. K., Tahir, N. M., Yusoff, Y. M., Thamrin, N. M., & Hasmi, A. H. (2023). Human detection from drone using you only look once (YOLOv5) for search and rescue operation. *Journal of Advanced Research in Applied Sciences and Engineering Technology*, *30*(3), 222–235.

Zhang, X., Feng, Y., Wang, N., Lu, G., & Mei, S. (2025). Aerial Person Detection for Search an d Rescue: Survey and Benchmarks. *Journal of Remote Sensing, 2025*, 0474.

Zhang, C., Patras, P., & Haddadi, H. (2019). Deep learning in mobile and wireless networking: A survey. *IEEE Communications Surveys and Tutorials*, *21*(3), 2224–2287. DOI: 10.1109/COMST.2019.2904897

Zhang, X., Liu, J., Chen, K., Gong, P., Liu, Y., & Wu, Z. (2024). Learning dynamic interactions and long-term patterns with spatio-temporal graphs for multi-vessel trajectory prediction. *IEEE Transactions on Intelligent Vehicles*, *9*(12), 7765–7780. Advance online publication. DOI: 10.1109/TIV.2024.3401864

Zhang, Y., Yin, Y., & Shao, Z. (2023). An enhanced target detection algorithm for maritime search and rescue based on aerial images. *Remote Sensing (Basel)*, *15*(19), 4818. DOI: 10.3390/rs15194818

Zhao, B., Zhou, Y., Song, R., Yu, L., Zhang, X., & Liu, J. (2024). Modular YOLOv8 optimization for real-time UAV maritime rescue object detection. *Scientific Reports*, *14*(1), 24492. DOI: 10.1038/s41598-024-75807-1

Zhao, J., Shi, W., & Xu, M. (2019). The potential of autonomous ships in green shipping. *Ocean Engineering*, *189*, 106299. DOI: 10.1016/j.oceaneng.2019.106299

Zhao, L., Song, Y., Zhang, C., Liu, Y., Wang, P., Lin, T., Deng, M., & Li, H. (2019). T-GCN: A temporal graph convolutional network for traffic prediction. *IEEE Transactions on Intelligent Transportation Systems*, *21*(9), 3848–3858. DOI: 10.1109/TITS.2019.2935152

Zhao, Z., Chen, W., Wu, X., Chen, P. C., & Liu, J. (2017). LSTM network: A deep learning approach for short-term traffic forecast. *IET Intelligent Transport Systems*, *11*(2), 68–75. DOI: 10.1049/iet-its.2016.0208

Zhou, Q., Liu, C., & Wu, Y. (2020). Predictive maintenance for ship propulsion systems using machine learning. *Journal of Marine Engineering & Technology*, *19*(5), 245–255. DOI: 10.1080/20464177.2020.1783002

Zhou, X., Zhang, L., & Qiu, Z. (2020). Intelligent navigation system for autonomous ships based on data fusion and deep learning. *Applied Ocean Research*, *95*, 102020. DOI: 10.1016/j.apor.2019.102020

Zhou, Z., Chen, X., Li, E., Zeng, L., Luo, K., & Zhang, J. (2019). Edge intelligence: Paving the last mile of artificial intelligence with edge computing. *Proceedings of the IEEE*, *107*(8), 1738–1762. DOI: 10.1109/JPROC.2019.2918951

Zhu, H., Wang, J., & Gao, R. (2025). EEG-based inverse reinforcement learning for safety- oriented global path planning in dynamic environments. *Applied Sciences (Basel, Switzerland)*, *15*(11), 6163. DOI: 10.3390/app15116163

Zhu, M., Zhang, W., & Xu, C. (2024). Ethical governance and implementation paths for global marine science data sharing. *Frontiers in Marine Science*, *11*, 1421252. DOI: 10.3389/fmars.2024.1421252

Zhu, Q., Xi, Y., Weng, J., Han, B., Hu, S., & Ge, Y. E. (2024). Intelligent ship collision avoidance in maritime field: A bibliometric and systematic review. *Expert Systems with Applications*, *252*, 124148. DOI: 10.1016/j.eswa.2024.124148

Zissis, D., Chatzikokolakis, K., Spiliopoulos, G., & Vodas, M. (2020). A distributed spatial method for modeling maritime routes. *IEEE Access : Practical Innovations, Open Solutions*, *8*, 1–1. DOI: 10.1109/ACCESS.2020.2979612

Zissis, D., & Lekkas, D. (2012). Addressing cloud computing security issues. *Future Generation Computer Systems*, *28*(3), 583–592. DOI: 10.1016/j.future.2010.12.006

Zissis, D., Xidias, E. K., & Lekkas, D. (2015). A cloud based architecture capable of perceiving and predicting multiple vessel behaviour. *Applied Soft Computing*, *35*, 652–661. DOI: 10.1016/j.asoc.2015.07.002

Zissis, D., Xidias, E. K., & Lekkas, D. (2016). Real-time vessel behavior prediction. *Evolving Systems*, *7*(1), 29–40. DOI: 10.1007/s12530-015-9133-5

About the Contributors

Froilan D. Mobo is a distinguished educator, researcher, and academic leader with extensive contributions to social studies education, curriculum development, and maritime studies. He currently serves as the Director of Research at Englishers International and is a Full Professor II at the Philippine Merchant Marine Academy (PMMA), where he teaches Demography and Economics in Social Studies. Dr. Mobo is actively engaged in scholarly research, with high-impact publications in social sciences, maritime studies, artificial intelligence, and human rights. His research works are indexed in Scopus and copyrighted at the Intellectual Property Office of the Philippines (IPOPHIL). He is also a Managing Director of FSH-PH Publications, which is visible in EBSCO. An advocate for academic excellence, Dr. Mobo is a regular member of the National Research Council of the Philippines (NRCP) under DOST and a former CHED RQAT Evaluator for ITE Education. He is also the Student Council President of Bicol University Open University (BUOU), where he is completing his third master's degree in Social Studies Education. Dr. Mobo has been a speaker and panel evaluator at national and international conferences, addressing topics such as academic writing, peer review processes, and Scopus-indexed journal publication strategies. He is a human rights advocate under the International Human Rights Movement (IHRM) Philippines and has participated in various academic steering committees, including the 2nd Education Research and Applied Business Conference (ERABC) 2024. His academic credentials include double doctorate degrees and a soon-to-be-completed third master's degree. His work continues to shape the fields of education, research, and scholarly publication both locally and internationally.

***

G Prasad is currently working as faculty in the department of Aerospace Engineering Chandigarh University, India. He has 9 years of experience in Teaching and research. He received his Bachelor of Engineering and Master of Engineering degrees with first class and distinction and Doctor of Philosophy in the field of Aeronautical Engineering from the Anna University, Chennai, India. Prof. Prasad G is the author of over 30 technical publications, 1 Patent Published and 1 Book published. His research interests include Unmanned Aerial Vehicles, Aircraft Structures, Computational Fluid Dynamics and Interdisciplinary research. He has completed three funded project sponsored by The Institution of Engineers (India) and Tamilnadu State Council for Science and Technology. Awarded Indian National Science Academy (INSA) Visiting Scientist Programme 2019 and Awarded Science Academies' Summer Research Fellowship Programme (SRPF) 2019. Further he is SPOC of NPTEL- IIT Madras, SPOC of ISRO IIRS outreach coordinator and SPOC for NASA International SpaceApps Challenge. Further key positions such as coordinator of Board of Studies, Outcome Based Education, Project Coordinator, Research Coordinator

Aadya Gupta is a dynamic aerospace engineering student eager to apply her academic knowledge and hands-on experience to contribute to the development of cutting-edge aerospace technology with a strong interest in aerodynamics, avionics, flight mechanics and unmanned aerial vehicles.

S. Ida Evangeline, presently working as an Assistant Professor in the Department of Electrical and Electronics Engineering at the Government College of Engineering, Tirunelveli, Tamilnadu, India. She obtained her Bachelor's in Electrical and Electronics Engineering from Manonmaniam Sundaranar University in 2003, and Master's in Power Systems Engineering from Anna University, Chennai, Tamilnadu, India in 2005. She completed her PhD degree in 2023 at Anna University, Chennai, Tamilnadu, India. She has published six research papers in international journals. Her current research interests include optimal power flow, renewable energy, soft computing, machine learning, and deep learning.

Snehal Malik is an undergraduate student in Aerospace Engineering at Chandigarh University. Her interests lie in aerodynamics, unmanned aerial systems and computational fluid dynamics. She is actively engaged in academic research and continues to explore new aerospace innovations.

Wes Harven Guillemer Maravilla is an EFL/ESL specialist creating innovative lessons to benefit students' English Language Learning throughout Asia since 2019 He is a Doctor of Philosophy in Education major in English Language Teaching graduate at the University of the Visayas. His research interests are writing instruction with reading literacy to improve reading comprehension among Taiwanese EFL Senior High School students. Mr. Maravilla has worked in various educational settings. He has working as a Foreign English Teacher with the Ministry of Education in Taiwan for the last two years and has taught high schoolers preparedness for college. Before his current role, he was a teacher at Malay College, teaching English to university students in Hospitality and Tourism Management. He also got a lot of experience with the Thai Ministry of Education where he taught a range of subjects to students in different levels. Mr. Maravilla believes in the Communicative Language and Project-Based Approaches. He is devoted to finding creative pedagogical approaches and enabling technologies that will help develop language learning, accessibility, and a sense of inclusion.

Mahira Mohamed Mowjoon earned a PhD in Engineering from the University of Technology Sydney, Australia and a Master's degree in Information Technology from the University of Sydney, Australia. She is currently a Lecturer in the Faculty of Engineering and IT at the University of Technology Sydney. Dr. Mowjoon brings a multidisciplinary background in engineering, information technology, and information systems, which enables her to actively contribute to both teaching and research across these domains. Her academic expertise is further complemented by extensive industry experience in a variety of roles. She is deeply committed to advancing student learning and remains actively engaged in impactful research.

Shekhar. R is a doctorate in Computer Science from MS University, Tirunelveli, a Postgraduate in Computer Networks and graduate in Computer Science and Engineering from the University of Bangalore. He has professional certifications in Microsoft and Cisco Networking. Dr.Shekhar joined Alliance in 2012 and has since been actively engaged in the areas of teaching, research, training and development. Dr.Shekhar has a vast experience of nearly two decades in teaching and has held many prestigious positions in various reputed institutions. He is a highly versatile teacher in various computer related subjects such as Data Mining, Image Processing, Computer Networks, Artificial Intelligence and many other subjects related to Computer Science. He has guided many postgraduate students in preparing dissertations in various subjects. He has presented papers in national and international forums, published number of articles in journals having high impact factors, and chaired many conferences. Dr.Shekhar has organized various workshops and conferences for highly reputed institutions. He is a lifetime member of ISTE and also holds the membership of various reputed international organizations. He serves as an editorial member of IJICTS, an International journal and has been a reviewer for several national and international journals. He is also a sitting member of different national and international associations in the field of Computer Science

Jayabhaduri Radhakrishnan is a Professor and Program Coordinator in the Department of Information Technology, Alliance School of Advanced Computing, Alliance University, Bengaluru, Karnataka, India. She holds a B.E. in Computer Science and Engineering, an M.E. in Computer Science and Engineering, and a Doctor of Philosophy (Ph.D.) in Computer Science and Engineering. She has successfully completed a Consultancy Project, an Externally Funded Project titled Niral Thiruvizha - Tamil Nadu's Grand Skilling Innovation Challenge, Anna University, Chennai, and five Intramural Funded Projects. She has authored and presented over 30 research papers in reputed international conferences and journals. As a reviewer for several scholarly venues, she actively contributes to the advancement of research in Artificial Intelligence and related fields. She is also a Life Member of several prestigious professional bodies, including the Institute of Electrical and Electronics Engineers, the Indian Society for Technical Education, and the Soft Computing Research Society.

Emi Sekikawa is a cybersecurity researcher and recent graduate of the Master of Information Technology (Extension) program at the University of Technology Sydney (UTS), Australia. She also holds a Bachelor's degree in Computer Science from a university in Japan. Her academic focus includes anomaly detection in IoT systems, AI-based intrusion detection, and UAV security protocols. During her studies, she conducted multiple research projects involving MAVLink telemetry analysis, Autoencoder-based anomaly detection, and network traffic simulation. Her work integrates machine learning with cybersecurity policy and protocol design. She is currently preparing for doctoral studies in Japan.

C.V. Suresh Babu is a pioneer in content development. A true entrepreneur, he founded Anniyappa Publications, a company that is highly active in publishing books related to Computer Science and Management. Dr. C.V. Suresh Babu has also ventured into SB Institute, a center for knowledge transfer. He holds a Ph.D. in Engineering Education from the National Institute of Technical Teachers Training & Research in Chennai, along with seven master's degrees in various disciplines such as Engineering, Computer Applications, Management, Commerce, Economics, Psychology, Law, and Education. Additionally, he has UGC-NET/SET qualifications in the fields of Computer Science, Management, Commerce, and Education. Currently, Dr. C.V. Suresh Babu is a Professor in the Department of Information Technology at the School of Computing Science, Hindustan Institute of Technology and Science (Hindustan University) in Padur, Chennai, Tamil Nadu, India.

Shuaimin Zhang is currently a candidate of the Master of Information Technology (Software Development) program at the University of Technology

Sydney (UTS). He also holds a Bachelor's degree in Finance, with an undergraduate research interest in microeconomics, macroeconomics, and monetary theory. His bachelor thesis explored cybersecurity risk insurance in China. During his postgraduate studies at UTS, his academic focus shifted toward software development, networking (including CCNA and CCNP-level knowledge), and cybersecurity.

Index

P

R

S

T

U

Y

www.ingramcontent.com/pod-product-compliance
Lightning Source LLC
LaVergne TN
LVHW081526300926
849161LV00006B/1318